Nanoreactor Engineering for Life Sciences and Medicine

Artech House Series
Engineering in Medicine & Biology

Series Editors

Martin L. Yarmush, Harvard Medical School
Christopher J. James, University of Southampton

Advanced Methods and Tools for ECG Data Analysis, Gari D. Clifford, Francisco Azuaje, and Patrick E. McSharry, editors

Advances in Photodynamic Therapy: Basic, Translational, and Clinical, Michael Hamblin and Pawel Mroz, editors

Biological Database Modeling, JakeChen and Amandeep S. Sidhu, editors

Biomedical Informaticsin Translational Research, Hai Hu, Michael Liebman, and Richard Mural

Biomedical Surfaces, Jeremy Ramsden

Genome Sequencing Technology and Algorithms, Sun Kim, Haixu Tang, and Elaine R. Mardis, editors

Inorganic Nanoprobes for Biological Sensing and Imaging, Hedi Mattoussi and Jinwoo Cheon, editors

Intelligent Systems Modeling and Decision Support in Bioengineering, Mahdi Mahfouf

Life Science Automation Fundamentals and Applications, Mingjun Zhang, Bradley Nelson, and Robin Felder, editors

Microscopic ImageAnalysis for Life Science Applications, Jens Rittscher, Stephen T. C. Wong, and Raghu Machiraju, editors

Nanoreactor Engineering for Life Sciences and Medicine, Agnes Ostafin and Katharina Landfester, editors

Next Generation Artificial Vision Systems: Reverse Engineering the Human Visual System, Maria Petrou and Anil Bharath,editors

Systems Bioinformatics: An Engineering Case-Based Approach, Gil Alterovitz and Marco F. Ramoni, editors

Systems Engineering Approach to Medical Automation, Robin Felder.

Translational Approaches in Tissue Engineering and Regenerative Medicine, Jeremy Mao, Gordana Vunjak-Novakovic, Antonios G. Mikos, and Anthony Atala, editors

Nanoreactor Engineering for Life Sciences and Medicine

Agnes Ostafin
Katharina Landfester

Editors

ARTECH
HOUSE
BOSTON | LONDON
artechhouse.com

Library of Congress Cataloging-in-Publication Data
A catalog record for this book is available from the U. S. Library of Congress.

British Library Cataloguing in Publication Data
A catalogue record for this book is available from the British Library.

ISBN: 978-1-59693-158-9

Cover design by Yekatarina Ratner

685 Canton Street
Norwood, MA 02062

10 9 8 7 6 5 4 3 2 1

Contents

1

Introduction to Nanoreactor Technology

Yen-Chi Chen, Qiang Wang, Agnes Ostafin

1.1 What is a Nanoreactor?

A nanoreactor is a nanosized container for chemical reactions. Unlike bench-top reactors or microreactors, the reaction space inside a nanoreactor strongly influences the movement and interactions among the molecules inside. As a result, the nanoreactor is not simply a holding vessel, but is a critical part of the chemical process. While nanoreactors are a relatively new material in science and engineering, many natural processes utilize nanoreactors. Some examples of these include cellular organelles and a variety of other organized biological microphases whose clearly distinguishable structures support a cascade of complex biochemical reactions. These places include the nucleus, mitochondria, Golgi apparatus, lysosomes, mitotic bundle, and the pores of channel proteins. There, the local concentrations and arrangements of molecules and ions are nonrandom, and this has profound consequences on chemical and photochemical processes that may take place inside.

The kinetics and mechanisms of chemical reactions in small-scale restricted geometries has been studied in micelles and vesicles [1], microfluidic devices [2], polymer and zeolite pore structures [3], and cells [4]. Considering an ensemble of nanoreactors, the reaction kinetics found in restricted geometries are different compared with the same reactions in bulk solvent, and they are hard to predict. First of all, for spaces containing a discrete number of molecules, the continuum approximation is no longer appropriate for describing the system. Relatively large fluctuations in the number of reagents per nanoreactor lead to very different kinetics and sometimes even reaction mechanisms among nanoreactors. One consequence of this is that the average behavior of the ensemble is not the same as would be the case for solution measurements. Second, the very large wall-area-to-volume ratio (the wall facing the interior of the nanoreactor) means that the frequency and type of interactions between molecules enclosed in the space may be influenced by the properties of the wall and reactant-wall interactions. These influences may result in molecular alignments, changes in molecular rotational dynamics (slows down or speeds up), and alteration in the mechanisms and rates of molecular relaxation.

Because the nanoreactor contains a finite number of molecules, the net yield of reaction may also be different from what is expected in the solution. Instead of the deterministic mean rate which is determined by the average frequency of the collisions in a system with large numbers of molecules, reactions of molecules distributed throughout an ensemble of nanoreactors are a probability phenomenon. Stochastic approaches have to be used to model the statistical fluctuations of the reactions between molecules [17]. For instance, the observed reaction kinetics of the system is an average of the kinetics of all the small systems that independently contribute to the overall kinetics. Each may have a different ensemble of factors that influence the reaction kinetics and mechanisms. This reaction rate is called the stochastic mean rate.

For a first-order reaction ($A \rightarrow B$), the deterministic mean rate and the stochastic mean rate are the same. However, for a second-order reaction ($A + A \rightarrow B$), the deterministic reaction kinetics are described by:

$$\frac{[A(t)]}{[A_0]} = \frac{1}{1 + [A_0]kt} \tag{1.1}$$

Assuming the reactant molecules in the small systems follow the Poisson distribution, the stochastic mean rate is [18, 19]:

$$\frac{\overline{N}(t)}{\overline{N}_0} = \sum_{n=1}^{\infty} B_n \exp\left[-\frac{1}{2}n(n-1)kt\right] \tag{1.2}$$

where:

$$B_n = \frac{2n-1}{2^n}\frac{e^{-\overline{N}_0}}{\overline{N}_0}\sum_{j=n}^{\infty}\frac{\overline{N}_0{}^j}{(j-n)!}\frac{\Gamma\left(\dfrac{j-n+1}{2}\right)}{\Gamma\left(\dfrac{j+n+1}{2}\right)}$$

and where $\overline{N}_0$ is the average number of reactant molecules in small system at initial time, $\overline{N}(t)$ is the average reactant molecules left after time *t*, and *k* is the reaction constant.

The difference between deterministic and stochastic reaction kinetics for a second-order reaction is more apparent for small average number of molecules. In deterministic reaction kinetics, all the reactants in an irreversible second-order reaction after infinite reaction time will be eventually consumed. However, in stochastic reactions, since molecules react in a pairwise fashion, half of the systems that contain an odd number of molecules will have one molecule left after completion of the reaction. To illustrate quantitatively, in an ensemble of nanoreactors filled with 7 molecules on average, up to 7% of the molecules will remain, and for one containing 3 molecules on average, up to 17% of the molecules will remain.

If the surface-to-volume ratio is very large, it means surface effects on the reaction kinetics cannot be neglected. If the concentration of reactants is high inside a nanoreactor, then the reaction rate can be increased since their mean free path within the nanoreactor is shortened by the existence of wall surfaces. This surface may repel the molecules generating more frequent collisions with molecules than would be expected from the same number of molecules in the same volume, minus the walls. The way the reactant interacts with the inner surface of the nanoreactor will affect the reactant's redox potential and Gibb's free energy, changing its reactivity. Interactions can influence the formation and evolution of the reaction transition state. The transition state for a bimolecular reaction is a highly excited intermediate state which must be formed before product can be formed. If the nanoreactor space is restrictive, then the two molecules may not be able to align themselves adequately to achieve this state, or to relax fully once it is formed, changing the product yields. Similarly, adhesion of the reactant at the interface can have a similar effect.

Strong absorption of reactants on surfaces slows down diffusion which thus additionally affects the reaction rate of reactants and coreactants. In the quenching of pyrene fluorescence by molecular oxygen, where both molecules are absorbed on a silicate surface, the quenching rate for pyrene is only ~40% of that in solution because both molecules need to diffuse to each other. Pyrene's quenching is greater on SiO_2 surfaces than it is on NaCl due to faster surface diffusion rates [5]. For nonadsorbed reactants, even periodic collisions with the walls of the nanoreactor can still slow down molecular motion and the diffusivity of molecules. This type of diffusion is known as Knudsen diffusion [6]. The transition state of the reaction pair also experiences this type of diffusion, and so can lose energy during the interaction, in some cases speeding up the reactions, and in others circumventing them.

Finally, the nanoreactor space could induce segregation or phase separation of the solvents and reactants inside influencing the reaction kinetics. For example, polar or aromatic solvents such as methanol and benzene in silica pores displace the absorbed pyrene on pore surface, decreasing the availability of solvent in the confined space, and increasing the concentration of pyrene in the solution phase. The amount of solvent adsorbed in systems within 4-nm pored silica was in the range 4.1×10^{-3} to 5.7×10^{-4} mol g^{-1} silica and led to a concentration change on the order of 10% [8].

To characterize molecular locations in nanoreactors experimentally requires good knowledge of the average locations of a molecule inside the nanoreactor. Spectroscopic methods are very popular, since they allow for relatively remote detection from outside the nanoreactor confines. However, the problem is that the spectral properties of the encapsulated molecule could be altered by other molecules within the nanoreactor environment and not just their location. The light-emitting excited state of the molecule can be influenced by the presence of many closely located dipoles in the nanoreactor. Depending on the duration of interaction, the effects on emission yields may be significant. For example, it has been shown that the interaction of arenes with charge transfer sites on SiO_2 surfaces decreases both the fluorescence yield and decay kinetics lifetime [7].

Of all the effects discussed above, which one will be the dominant effect is decided by the dimension of the confined space, the number of molecules in each confined space, and the interaction between the wall and the reactants. In general, as the dimensions become smaller, and fewer reacting molecules in each space increasingly interact with the surface and each other, the difference between reaction kinetics in confined space and in bulk will be larger.

1.2 Examples of Nanoreactor Systems

1.2.1 Overview

Recent years have seen the emergence of a rich array of natural and synthetic structures which are capable of nanoreactor function. These include a wide variety of polymeric and lipid hollow spheres, biomineralized membranes, and cells. Many of these materials are being developed for use in the preparation of other types of nanoparticles, to improve the efficiency of chemical processing, as stand-alone or implantable smart drug delivery vehicles, as nanomedicines, as biosensors, and as replacement tissues. Their development has been enabled by significant improvements in the ability of chemists to control nanostructure geometry and properties. Thus, a significant portion of recent scientific research has focused on the chemistry and physics of these materials, and how this may affect optimization of their internal properties and effect on reactions. Several reviews of nanoreactor systems have been published over the last few years and the following sections are a survey of nanoreactor types available. Some of these already have shown to have clear applicability to life science and medicine, while others have potential in these fields but their development to date has emphasized other technology areas.

The nanoreactor concept first emerged in the late 1990s, and several early reviews point to its potential in chemical transformations and medicine [9–11]. Since then, other reviews highlighting the synthesis and general characterization of specific categories of nanoreactors have been published. These include self-assembled nanoreactors [12, 13], nanoreactors and nanocontainers [14], biomineralized nanoreactors [15], planar, inorganic, polymeric [16], and composite nanostructures with nanoreactor-like porosity [17], amphilic block copolymer nanoreactors [18], and polyelectrolyte nanoreactors [19]. In general, inorganic nanoreactor structures have been of interest for high-temperature, high-pressure reactions of industrial importance since the inorganic matrix is mechanically and chemically strong, and so are able to withstand extreme conditions of industrial processes. In contrast, self-assembling organic structures have much broader applicability and are used to template the synthesis of other nanostructures as well as forming chemical reservoirs for drugs, chromophores, and other reagents.

Molecular organic and biomacromolecular nanoreactors are the smallest organic nanoreactor structures composed of one, or a few large molecules that are assembled so that they form a hollow space into which can fit at least one other molecule. The entrapped molecule can serve as a reactant, and the efficiency and nature of the reaction it may undergo, can be changed from what it would be in solution. The pocket in which the reactant resides can change the electronic distribution or impart strain in the inserted molecule, facilitating subsequent chemical transformations.

Porous macroscopic solids such as silicates and other metal oxide frameworks have long been recognized to have unique impact on chemical reactions that occur inside their pores. Their pore spaces are considered as an interconnected network of nanoreactors. Such nanoreactors are synthesized using a top-down strategy and their properties are largely limited by the composition of the matrix material and any residual porogenic substance used in their formation. Postsynthesis modification of the nanoreactor spaces is possible, although, if the size of the monolith is significant, uniformity of treatment throughout may be difficult to achieve.

Micelles and vesicles are much larger organic nanoreactor structures comprised of thousands, to tens of thousands of lipid, surfactant, or short-chain polymeric molecules which spontaneously self-assemble into closed structures. The size, shape, and surface chemistry of the structures obtained depends on the charge and hydrophobicity of different parts of these molecules, the solvent system in which they have formed, and the presence of other surfactants and lipids. Micelle and vesicle structures are relatively flexible and somewhat permanent making them good hosts for chemical reagents with hindering their accessibility. Therefore, they have been used as carriers to solubilize chemical substances and localize the occurrence of chemical reactions. The state of the art for this area is being advanced by the development of many nonnatural surfactant and lipid structures made from a variety of polymeric and block copolymeric materials. These offer similar self-assembling capability but with a wider array of chemical and physical characteristics that allows them to be used at elevated temperatures and pressures, and under chemically harsh conditions of pH, temperature, shear, and oxidative chemistry.

Recently, there has been much interest in the use of bacterial, viral, and mammalian cells as nanoreactors. For instance genetically engineered bacteria produce complex chemical products more efficiently than would a soluble enzyme. Another example is the virus *capsid* which can be emptied and used as a container for reactive substances. Using such structures takes advantage of the extensive material optimization that natural evolution has already performed, along with the rich array of molecular transporters that can be used to control the contents of the internal space. Understanding these complex structures also provides inspiration for the development of synthetic biomimetic structures.

1.2.2 Molecular Organic Nanoreactors

Molecular organic nanoreactors are generally large molecules or molecular complexes which take on a unique shape. A cavity inside this structure is externally accessible, and one or more molecules are able to enter and undergo chemical transformations. The caging nanoreactor or molecular basket as it is called in some instances, may, or may not, participate in these transformations directly, but its presence influences the outcome. As pointed out earlier in this chapter, a wide range of enzymatic structures both natural and synthetic could be included within this category of nanoreactor. Although these have clear biological or biomedical importance, a thorough treatment of these systems would be well beyond the scope of this text.

What effects molecular organic nanoreactors exert on chemical reactions depends on the nature of the structure and that of the reactants. For example uracilophanes are amphiphilic macrocycles that are made by combining several identical molecular pieces using a quaternary ammonium bonding. They are able to increase the yield of the hydrolysis of alkyl phosphonates up to 30-fold depending on the specific macrocycle structure [20]. Other examples include the enhanced methanolysis inside molecular baskets which is attributed to the ability of the basket to able to concentrate ethanol from a solution [21], the controlled phototransformation of stilbene in van der Waals nanocapsules [22], and the efficient cycloaddition of arene in a self-assembled nanocages [23]. Recently, very small molecular nanoreactors such as rhombibicubooctahedral nanocapsules 4 nm in diameter linked by 24-imine bonds capable of encapsulating tetralkylammonium salts in solvents like toluene for reaction [24], and pyrogallol 4 arene hexameric capsules have been reported (see Figure 1.1) [25].

1.2.3 Macromolecular Nanoreactors

For the purpose of this chapter, we will consider macromolecular nanoreactors to refer to structures with multiple repeating units. Given this broad definition, organic polymers, proteins, and carbonaceous materials will be considered in this section.

Organic polymer nanoreactors are particularly rich in terms of structural variety. Examples range from relatively simple polymer aggregates to block copolymers, polymerosome, dendrimers, polyelectrolyte-layered materials, and hydrogels. Organic polymer materials have been used as microreaction cages [26], enzymes [27, 28] for photochromic dyes [29], and other nanoparticles (see Figure 1.2) [30]. A clear advantage of organic polymer is that it is possible to molecularly imprint nanoreactors for example, for

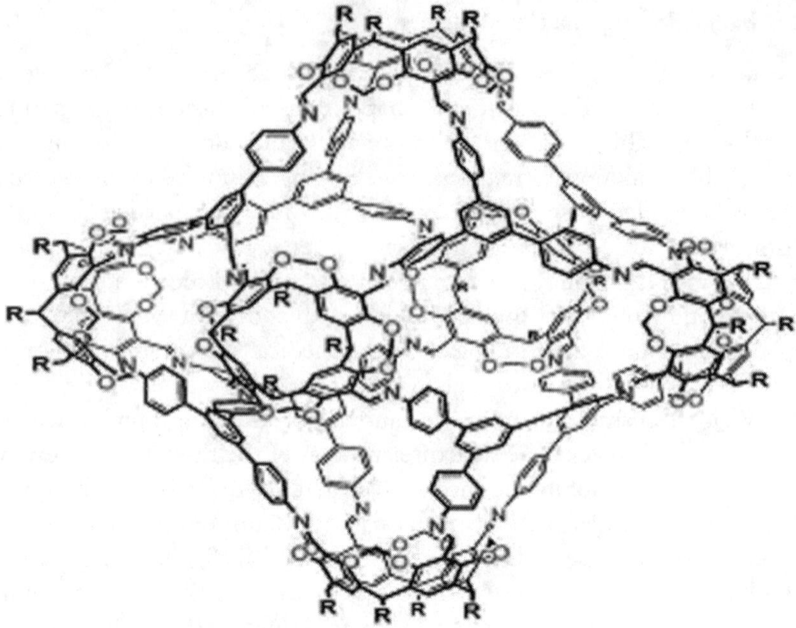

1 R = pentyl

Figure 1.1 Rhombicuboctahedron nanocapsules linked by 24-imine bonds capable of encapsulating tetralkylammonium salts in a solvent for reaction. Copyright Wiley–VCH. Reproduced by permission [24].

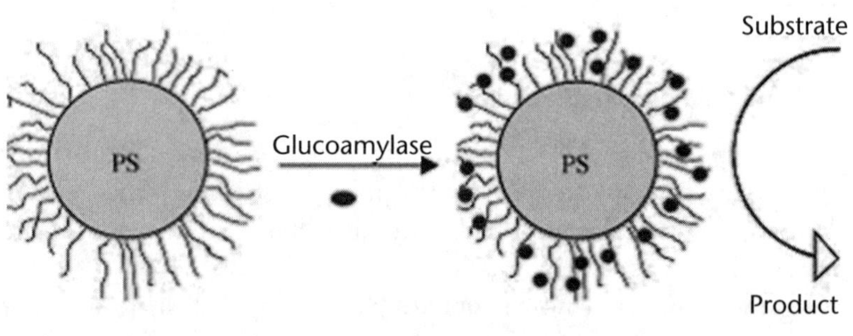

Figure 1.2 Schematic representation of the organic polymer nanoreactor used as a matrix to support the enzyme. The nanoreactor consists of a polystyrene core onto which long chains of poly(acrylic acid) (PAA) have been grafted. Glucoamylase (enzyme) adsorbs spontaneously from solution onto the spherical polyelectrolyte brushes if the ionic strength is low. Copyright Wiley–VCH. Reproduced by permission [28].

regioselective reactions [31], and to generate larger monoliths with nanoreactor capability [32].

While most organic polymers are capable of aggregation into small colloidal structures of relatively uniform size under appropriate solvent composition, ionic strength, and temperature, precise control over their three dimensional structure is not possible. It can, however, be achieved using designer block copolymers, which are short polymers consisting of two or more kinds of repeating units arranged nonrandomly in the polymer chain. By varying the number, spacing, and branching of these blocks within the polymer, it is possible to direct the way in which the polymer assembles and interacts with other molecules in the surroundings.

Block copolymer nanoreactors [33, 34] can form micelles, microemulsions, and polymerosomes, a polymer analog of liposomes. In many respects, the block copolymer can be considered to be a specialized surfactant that organizes its structure so that hydrophilic and hydrophobic domains are found at opposite ends or sides of the structure. These molecules are then free to interact with one another which can lead to self-assembly into closed structures if the change in Gibb's free energy reduction compensates for loss in entropy. The familiar packing factor concept can apply to these structures since they too may form cylindrical or cone-shaped molecules. However, since the surface and contact area between domains of adjacent macromolecules is much greater than for smaller surfactant molecules. This simple picture fails to adequately predict the structural richness of these materials.

Micellar and microemulsion structures made from block copolymers have been used with much success for the synthesis of metal and metal oxide nanoparticles and clusters [35–39]. These nanoparticles include: PbS [40], Au [41, 42] , Ag [43], CdS [44], doped ZnS [45, 46], as well as some oxide nanomaterials [47]. Depending on the structure of the block copolymer it is possible to generate nanoreactors with pH-dependent permeability [48], a variety of core solvents and materials (including proteins) [49–51], self-catalyzing nanoreactors for esterolysis [52], and nanoreactors which facilitate the hydrolytic cleavage of organic phosphonate have been reported (see Figure 1.3) [53].

Drug delivery [54] is another area where these nanoreactors are being explored. Rather than relying on conventional dissolution, disruption, or degradation of the carrier, it has been shown that it is possible to supplement some organic polymer nanoreactors with channel proteins to facilitate controlled material transport in and out of the nanoreactor [55, 56].

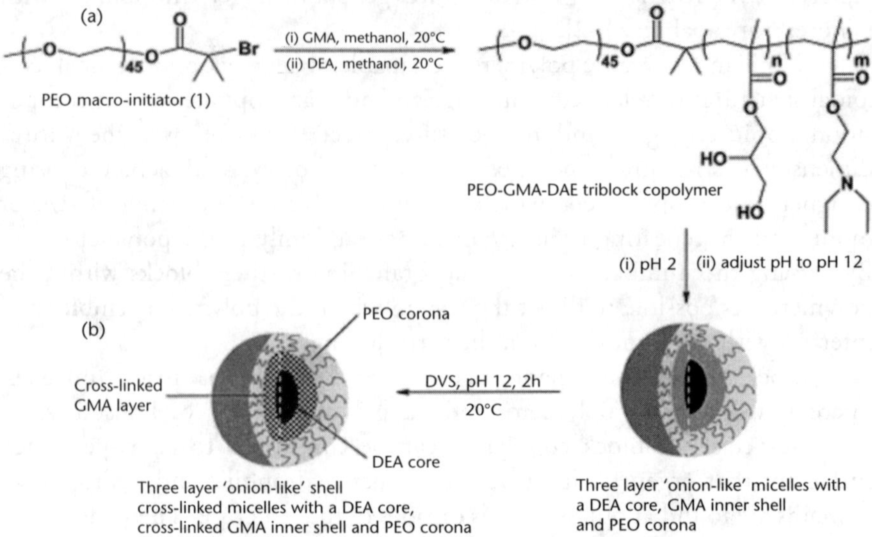

Figure 1.3 Block copolymer nanoreactors generated with pH permeability. (a) Reaction scheme for the synthesis of the PEO-GMA-DEA triblock copolymers; (b) schematic illustration of the formation of three-layer onionlike micelles and shell cross-linked micelles from PEO-GMA-DEA triblock copolymers. Copyright ACS. Reproduced by permission [48].

Ordering of organic polymer nanoreactors in two and three dimensions has also been explored since for most applications a macroscopic physical structure is convenient for handling [57, 58]. Nanoreactors have been formed by gaseous voids formed using supercritical CO_2 in block copolymer matrices [59], by cavitation [60], or tubular core shell microstructures in chiral diblocks [61]. In addition, there have been reported nanoreactors made from block copolymers that are able to open and close while attached to a surface [62], and which can create arrays of metal nanodots [63].

Polymerosomes are made from block copolymers capable of self-assembling into closed geometries entrapping a second material in the core space. This material could be solvent (e.g., water), solutions, and metals [64–67], semiconductor [68], and magnetic nanoparticles [69]. Perhaps the most relevant applications to biomedicine have taken place using enzymes and multilayered polymerosome structures. For example, the possibility of supporting cascade reactions of enzymes within polymerosomes was demonstrated

[70–72] as was the use of multilayered structures to form different reaction environments in the same particle (see Figure 1.4) [73, 74]. While not strictly a polymersome, it is possible to use the spaces created by a polymer brush as nanoreactors. This brush is covalently linked to a second larger nanoparticle for support [75, 76].

Amphiphilic or polyelectrolyte polymers [77, 78] formed by the sequential deposition of multiple layers of polymer material are used for the construction of pH, thermoresponsive [79], and charge-selective [80] nanoreactors. As with some of the other examples already mentioned, these have been used in the synthesis of Ag, Au, and various other nanoparticles (see Figure 1.5) [81–83]. Such nanoparticles can be used in catalysis applications, for instance Co metal cored ones are capable of catalyzed hydrolysis of epoxides with 99% yield [84]. Capsules made with embedded enzymes [85] and vesicles [86, 87] also have been reported.

Dendrimers [88] are large molecules with extremely well-defined structures that are nearly perfectly monodisperse. Dendrimers consist of three major architectural components, a unique multiple-branched core particle, branches, and end groups. They are formed by controlled hierarchical synthesis, which is a bottom-up approach, in which the multiple-branch core molecules act as a seed for the next layer or generation of constructed from assymetric branched polymers. The growth of dendrimers is self-limiting, and ends when the surface area of the terminal layer is maximally dense. Fundamental research in branched polymers is very extensive today and beyond the scope of this review, but their usefulness in the construction of

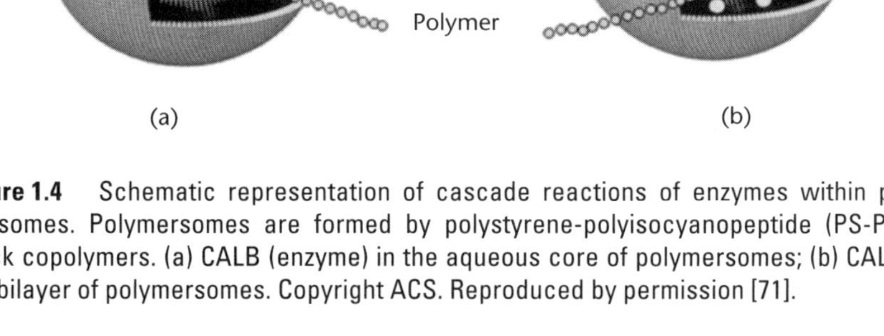

Figure 1.4 Schematic representation of cascade reactions of enzymes within polymersomes. Polymersomes are formed by polystyrene-polyisocyanopeptide (PS-PIAT) block copolymers. (a) CALB (enzyme) in the aqueous core of polymersomes; (b) CALB in the bilayer of polymersomes. Copyright ACS. Reproduced by permission [71].

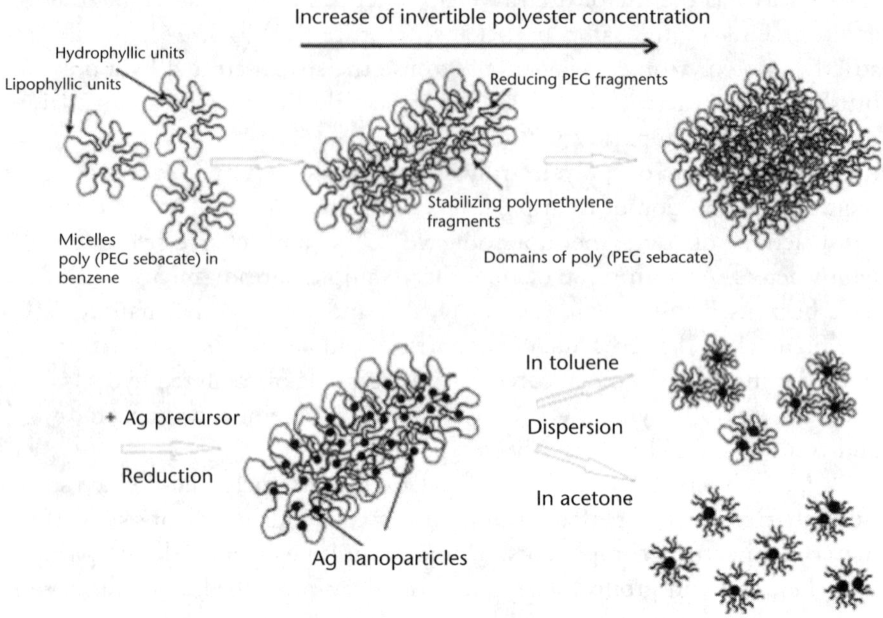

Figure 1.5 Schematic representation of silver nanoparticles synthesized in amphiphilic polyester nanoreactors. Copyright ACS. Reproduced by permission [81].

nanoreactors was recognized relatively early [101–103]. Dendrimers are highly versatile nanoreactors for enzymatic reactions [89, 90], and the synthesis of nanoparticles of CdS [91], Cu [92], Pd [93, 94] , Pt [95], Au [96], Ag [97]. Other application areas include sensors [98] and chemical catalysis [99, 100].

Hydrogels are water-saturated polymers, with generally excellent biocompatibility characteristics. Depending on the nature of cross-links used in the hydrogel, it can be made to cleave on trigger or over time, changing the porosity and elasticity of the matrix. For this reason hydrogels are used in the development of tissue-engineering scaffolds [104, 105], and in the metabolic byproducts of proliferation cells used to stimulate matrix degradation according to the evolving needs of the repairing tissue. This feature is also useful for drug delivery applications [106]. The pore spaces within the hydrogel are nanoreactors. In these spaces, just like in many of the other materials already discussed, it has been shown to be possible to produce metal [107–110] and metal oxide nanoparticles [111, 112]. The hydrogel structure defines the dimension and geometry of the void, water-filled pore spaces, nanoparticles of various shape and size and can be produced (see Figure 1.6) [113]. The antibacterial action of many metal nanoparticles such

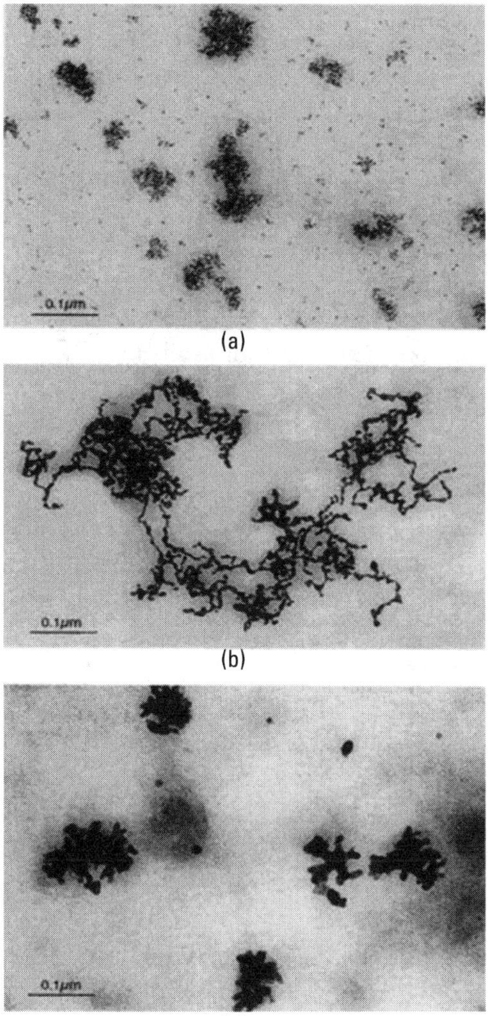

(c)

Figure 1.6 TEM image of various shape of nanoparticles synthesized with different hydrogel formulations and a reducing agent: (a) uncontrolled particle morphology arising from a fast reduction with multiple nucleation; (b) threadlike morphology after *slow* reduction with sodium borohydnde; (c) nuggetlike morphology after a reduction with hydrazine. Copyright Wiley–VCH. Reproduced by permission [113].

as silver is of particular biomedical importance and such composite hydrogel materials are being developed for use in antiburn dressings and bone replacements.

Carbon nanotubes are included in this discussion since they are a kind of organic macromolecule, constructed from numerous carbon atoms arranged in closely packed hexagonal format. The inner diameter of carbon nanotubes is small, of the order of several Ångstroms. The inner space may be filled with fluids like CO_2 [116] and used as a nanoreactor to nucleate smaller nano-objects [114, 115]. Carbon nanotubes have been used to produce metal nanopowders [117], Mg_3N_2 [118] and iron [119] nanowires, magnetite [120] and Gd_2O_3 nanoparticles (see Figure 1.7). The outer walls of the carbon nanotubes can be functionalized with charged groups, and these places used to bind metal catalysts for organic transformations [121]. In addition to tube geometries, carbon is also capable of yielding a variety of structures of varying size including multiwalled tubes, spheres, horns [122], onion-like structures [123], and branched configurations. These too can be used to synthesize nanoparticles and in some instances can generate super high internal temperatures (>2000° C) and pressures (>40 GPa).

It's worth noting that materials other than carbon could be used to generate tube nanoreactors. Some of these include transition and lanthanide metal oxides [124], organic polymers [125, 126], DNA [127], and proteins [128, 129]. Synthetic geometries of DNA form nanoreactors inside which Ag, CdS nanoparticles can be synthesized [130–132], and peptide nanodoughnuts self-assemble from peptides and gold salts, leaving gold nanoparticles inside following reduction [127] (see Figure 1.8).

Figure 1.7 (a) XRD pattern and, (b) SEM image of the Mg3N2 nanowires produced within carbon nanotubes. Copyright ACS. Reproduced by permission [118].

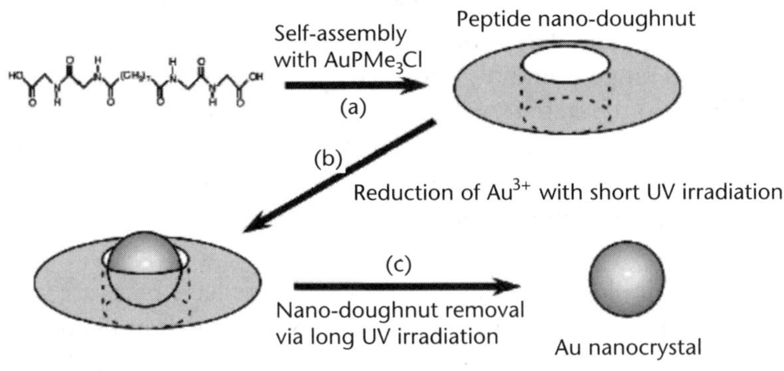

Figure 1.8 (a) Peptide nanodoughnut self-assemble from peptide and gold salts; (b) Au ions in the cavity are reduced by short UV irradiation (<20 min); (c) longer UV irradiation (>10 h) destroys the nanodoughnut to release the Au nanocrystal. Copyright ACS. Reproduced by permission [127].

Stereospecific reactions such as chiral center formation and pyramidal inversion were also found to be facilitated in a protein nanoreactor [133].

As with other macromolecules, protein nanoreactors can be constructed in a variety of forms ranging from a simple core surrounded by shell structures [134], molecular assemblies like nanosomes, which contain dockerin-engineered enzymes in chimeric scaffoldins for cellusome function [135], and three-dimensional architectures [136], including tubules and layered structures. The alpha-hemolysin pore can be used as a nanoreactor for the photoisomerization of azobenzene, which acts to stabilize the *cis* state to make more complete photoisomerization possible with less degradation [137]. Cellulose fibers [138], protein layers [139], and other protein assemblies [140] can be used as a nanoreactor to make noble metal and Ga_2O_3 nanoparticles since the high oxygen content in the underlying structure helps anchor metal ions into a nucleation site.

1.2.4 Micelle, Vesicles, and Nano/Micro/Mini Emulsions

Micelle, vesicles, and nano/micro/mini emulsions [141–143] made from small surfactant, lipid molecules, or polymeric molecules are another category of nanoreactors. The relatively well understood self-assembly dynamics of these materials, and the ease of formation of enclosed structures with specified size, and in some cases shape, is very attractive.

Micelles can either be oil droplets suspended in water or water droplets suspended in oil. The description droplet is a misnomer since the size of these structures range from only a few tens of nanometers to nearly a micron in size. Critical to the stability of these nanoreactors is the presence of a third molecule, either a surfactant or a lipid which can bridge the interface between the two phases, lowering the extremely high surface tension and making them stable for long periods of time, even under turbulent mixing conditions.

Regular micelle nanoreactors with an oil core have been used for condensation reactions with aldehyde [144], peroxidase catalysis of aniline polymerization [145] (see Figure 1.9) [146], Rh catalysis of hydrogenation reactions [147], photophysical events [148], and as lubricants [149]. Mixing with polyethylene glycol can be used to make long circulating nanoreactors that evade the RES system [150].

Reverse micelles [151] with water core are commonly employed in the synthesis of nanoparticles of Co [152, 153], ferrite [154], CdTe [155], CdS [156], gold [157], CeO_2 ZrO_2 [158], zincphosphonates [159], starch [160], PANI/TiO_2 [161], magnetitie [162], ferrihydrate [163], $SrTiO_3$, Sr_2TiO_4 $PbTiO_3$ [164] and phosphors [165]. Reverse micelles are more than just a holding vessel to template these reactions. Their shape [166] (see Figure 1.10) may be dynamic during reactions and so the mechanism by which they affect the product outcome may be more complex [167], than the simple solubilization of metal ions and clusters [168]. The effects on nanoreactor function are strongly dependent on the structure of the lipids/surfactants [169]. Other known functions for reverse micelle nanoreactors include scavenging of environmental toxins [170], prevention of gelation and improved size control [171], and elimination reactions of tertiary alkyliodides [172].

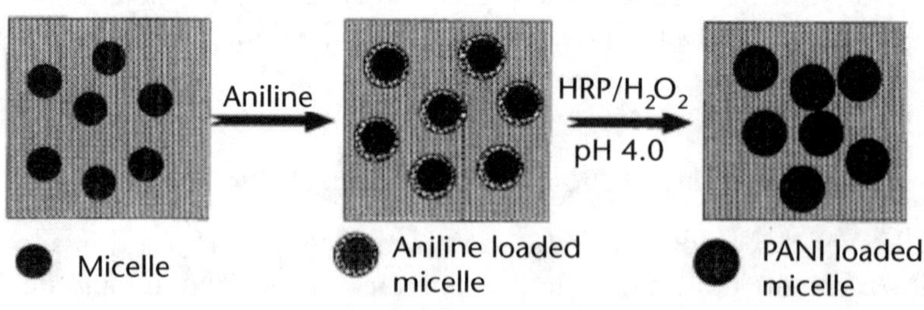

Figure 1.9 Peroxide catalysis of enzymatic polymerization of aniline in aqueous micelle solutions. Copyright ACS. Reproduced by permission [145].

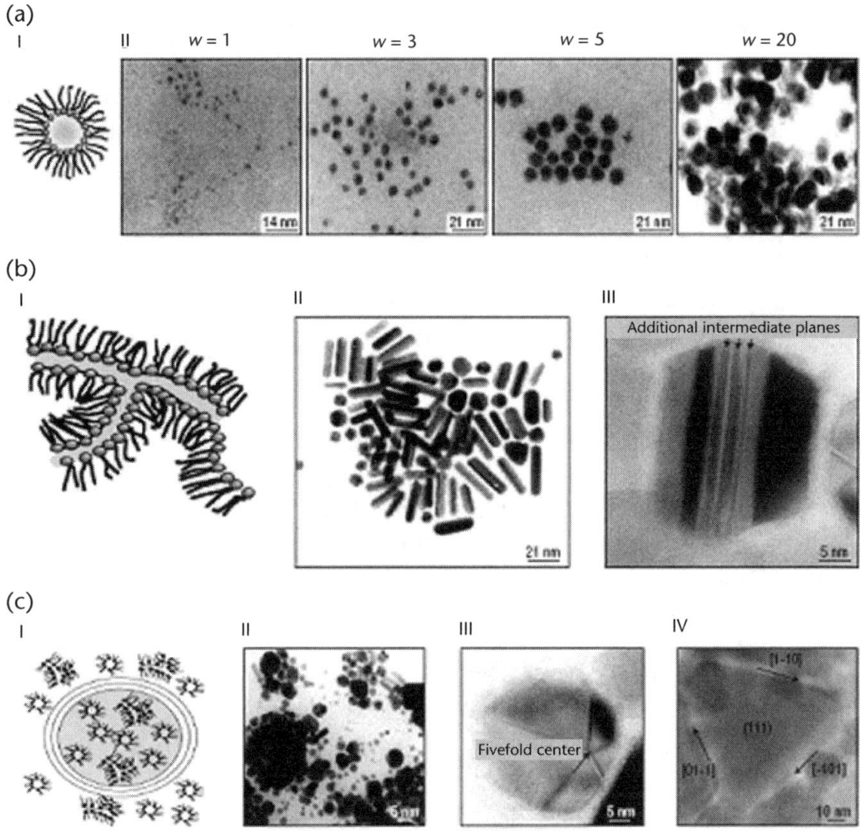

Figure 1.10 The different shapes of copper nanocrystals produced in different shape of self-assemblies of surfactant-H_2O-isoctane (reverse micelles) solution. (a) I. Reverse micelles. II. TEM image of the formation of nanocrystal with different size, *w*, which is controlled by the size of water-in-oil droplets. (b) I. Interconnected cylinders. II. TEM image of the produced spherical and cylindrical nanocrystals. III. cylindrical particle composed of a set of deformed f.c.c.tetrahedra bounded by (111) faces parallel to the fivefold axis with additional planes. (c) I. Supra-aggregates. II. TEM image of various nanocrystals. III. Particle composed of five deformed f.c.c.tetrahedrals bounded by (111) planes. IV. Large, flat nanocrystals [111] oriented and limited by (111) faces at the top,bottom, and edges. Copyright Nature Publishing. Reproduced by permission [166].

Liposomes and vesicles [173] are water-cored nanoreactors surrounded by a bilayer of lipids and in the case of natural vesicles, some proteins. The water-filled space can be filled with a variety of materials [174, 175] and decorated with channel proteins for improved mass transport [177–179]. The spaces between multilamellar vesicles can be used as nanoreactors to synthesize inbetween nanoparticles layers [176] (Figure 1.11).

Emulsions are larger two-phase systems. Reactants preferentially accumulate in the core phase or at the interface, where their increased concentration and favorable orientation can speed up the process [180, 181]. At very large sizes even though there is reagent concentrated at the periphery, the interface is so far removed from the materials inside the core region that the reaction is inhibited [182]. Nevertheless, emulsions are useful in the synthesis of nanoparticles including Ag/AgI [183], ZrO [184–187], ZnSe [188], ZnS [189–191], ZnO [192], SnO_2 [193], CdS/ZnS [194, 195], $BaTiO_3$ [196], BaZrOMeO [197], magnetite/silica core shell [198], Ag [199] semiconductors [200], and other inorganic nanoparticles [201]. Other uses have included making flavor delivery more efficient for food [202], to cleave phosphorous acid esters, about 1000-fold more efficiently [203], enantioselective enzymatic reactions [204], PCR [205, 206], and the production of polythiophene by Fe^{3+} oxidation in a thiophene nanoreactor in a surfactant droplet (O-W emulsion) [207] (Figure 1.12).

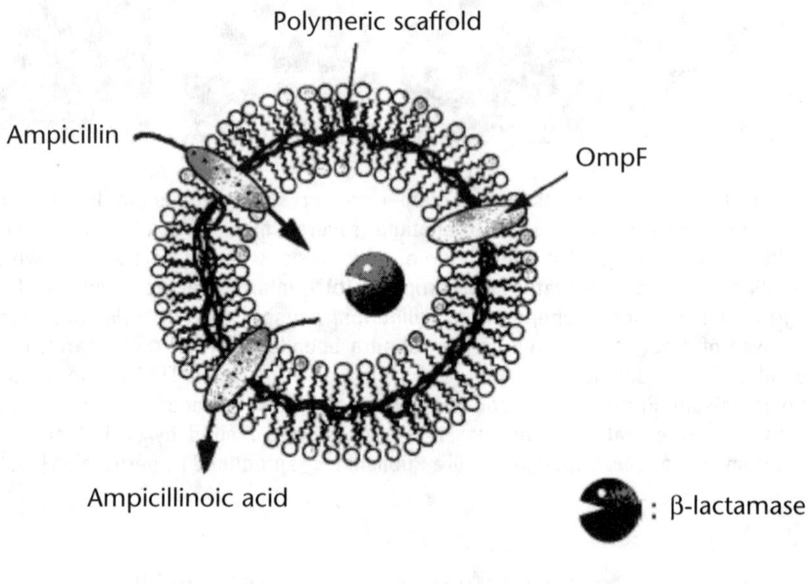

Figure 1.11 Schematic representation of enzyme (β-lactamase) encapsulated polymer-stabilized nanoreactor with decorated channel protein (OmpF, ampicillinoic acid, and ampicillin). Copyright ACS. Reproduced by permission [179].

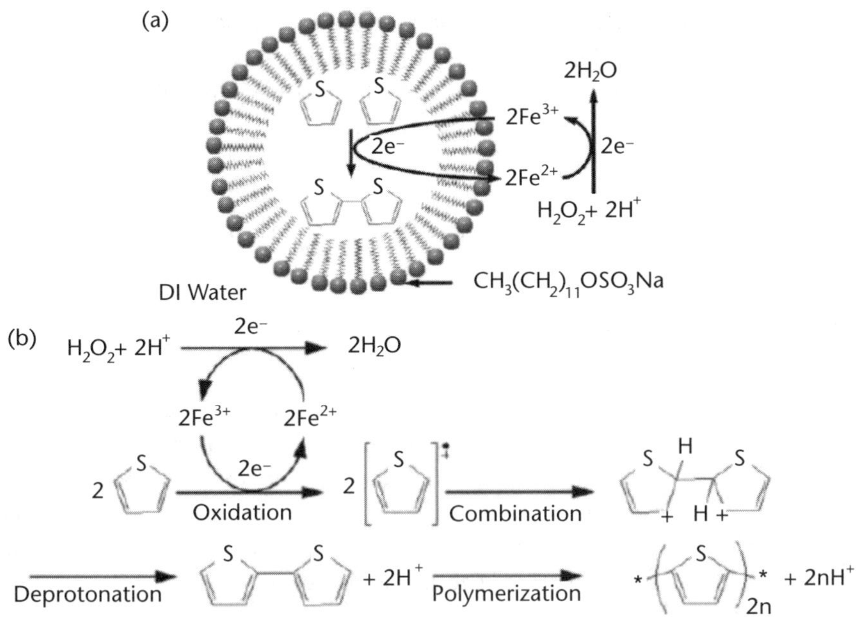

Figure 1.12 (a) Schematic representation of the mechanism for the production of polythiophene nanoparticles by Fe^{3+} oxidation in a thiophene nanoreactor in a surfactant droplet; b) Detailed reaction mechanism of Fe^{3+} oxidation and thiophene polymerization. Copyright Wiley periodicals. Reproduced by permission.

Figure 1.13 TEM image of Bi nanoparticles made by lyotropic liquid-crystal nanoreactors: (a) Hexagonal phase, (b) lamellar phase, and (c) inverse hexagonal phase. Copyright ACS. Reproduced by permission [212].

Liquid crystals, ionic liquid films, and lipid tubules [208–211], have also been shown to have utility as nanoreactors to make other nanoparticles (See Figure 1.13) [212–214]. Again, the structure of the liquid crystal phase controls the particles obtained and the size and property of the void space [215]. Reactions include Suzuki coupling [216] and the hydrolysis of phosphonates and phosphate [217]. The liquid crystal nature of the cellular plasma membrane, which coordinates spatial and temporal control of lipid metabolism, trafficking, and organization, [218] has similar capability [219].

1.2.5 Porous Macroscopic Solids

The last category of nanoreactors are those formed by the void spaces in larger monoliths. One of the examples are mesoporous silicates and zeolites [220–223]. The void spaces inside these materials can range from a few angstroms to nanometers in diameter depending on whether they we created via the incorporation of lipid, surfactants, or other large atoms and molecules. The pore structure can be further manipulated by adding molecules like cyclodextrin to make worm-like geometries [224] or via evaporation-induced self-assembly of porous silica with nanotextures (see Figure 1.14) [225]. Enhanced catalysis is achieved when surfactant is mixed with silica/ammonia molybdate catalysts. These later decorate the inner surface of the nanoreactor spaces throughout the material and can be accessed by other reactant solutions. Such designs have been used for nanoparticle formation [226], cyclohexene oxidation [227] to produce metal nanoparticles [228, 229], to produce oxide nanocrystals [230], to produce magnetic nanocomposites [231], to support enzymes [232], for epoxidation [233], for excited-state deprotonation [234], for oxidation of hydrocarbons [235], for halogen switch reactions [236], to form nanowires [237], for CNTs [238], and for the digestion of proteins for protein analysis [239, 240].

Reactions in the mesoporous oxide pores are different from what happens outside the pores [241–243]. Pores improve/speed the surface orientation of reagents like fluorine attached to 1, 3, diphenylpropane, and exert proximity effects on free-radical reactions [244]. The curvature of pores 1.6 to 2.8 nm affects H transfer to from radical intermediates [245]. Reagents can also migrate along the inner surface of the pore nanoreactors and be absorbed [246]. A consequence of this is that substitution reactions of metal carbonyls happen 103-times faster when in sodium zeolite Y nanoreactors, because binding to the inner surface of the pore affects orientations, transition states, and yields. Similarly nearly 100% efficient epoxidation of alkenes and 80% to 90% selectivity can be obtained [247]. In addition protein

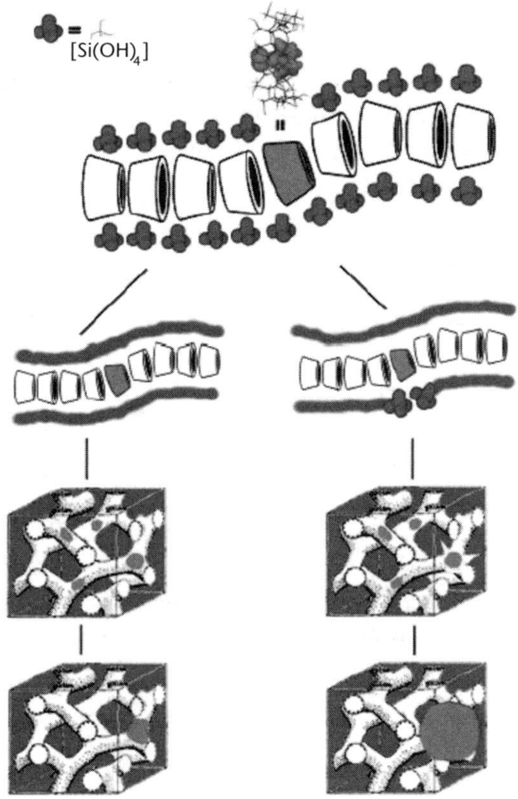

Figure 1.14 The worm-like geometries when cyclodextrin is added to the pore structure of mesoporous silicate. To demonstrate the dependence on the aging conditions, the silica walls are fully condensated before the metal particle nucleation and growth starts (left side), and after (right side). Copyright ACS. Reproduced by permission [244].

digestion is faster and more complete yielding better sequencing [248–250], and the oxidation of alkenes by $CoCl_2$ is more selective [251]. These materials have been used in the synthesis of nanocomposites [252, 253], production of photocatalysts [254], CdS [255], NbCo/Nb and other metal nanoparticles [256], for selective epoxidation [257, 258], for free-radical grafting of maleic anhydride onto polypropylene [259], and for photocatalysis [260].

Functional porous materials also include hollow inorganic nanoparticles formed through Ostwald ripening [261], halloysite and polyelectrolyte capsules [262], hollow TiO_2, calcium phosphate [263], monoliths [264], chemically tailored nanoreactors in silicate and zeolites [265], hollow silicate with tunable wall thickness [266, 267], and hybrid PEG/cyclic or cubic silica nanoparticles [268]. Such materials have found usefulness in lab on a chip monolith nanoreactors [269], and three-dimensional porous metal ions or metal-oxygen clusters networks, chains, and layers [270]. Layered double hydroxides consisting of clay-Mg-Al form ionic lamellar solids (anionic based) [271, 272]. The interstitial spaces can be used to make Pd [273], FePT:C [274, 275], gold [276], and magnetite [277] Au/Pt [278]. For clays adsorption of ions, its surface is key in the formation of nanoparticles [279], leads to the possibility for shape-selective chiral reactions [280], and generates a solid-phase coordinating environment to make nanoparticles of uniform size [281].

Other examples of inorganic nanoreactors include Keggin structures, [282, 283] which are nanopolyoxometalates such as phosphotungstinic acids [284], hallyosite tubes which have experienced biomineralization reactions inside [285], MoS_2 nanotubes [286], silica Xerogel [287], and lithographically etched nanopores on quartz useful in biochemical sensing potential [288]. Once a material product is formed inside, the matrix itself can be sacrificed using nanoscale explosions to release the contents (see Figure 1.15) [289, 290].

1.3 Conclusions

As can be seen from this short review, nanoreactors of various composition are ubiquitous in technology and material science. Their potential for

(a) (b)

Figure 1.15 TEM image of MoS_2 nanotubes: a) the MoS_2 nanotubes with encapsulated MoS_2 nanoparticles; b) single MoS_2 nanoparticle and their aggregates inside a thin-walled MoS_2 nanotube. Copyright Wiley–VCH. Reproduced by permission [286].

controlling chemical processes opens the door for many applications in biomedicine ranging from sensors, drug delivery, and medical devices. The subsequent chapters will provide a more in depth view of some of these technologies most relevant to biomedical areas.

References

[1] Karlsson M., Davidson M., Karlsson, R., Karlsson, A., Bergenholtz, J., Konkoli, Z., Jesorka, A., Lobovkina, T., Hurtig, J., Voinova, M., Orwar, O., "Biomimetic Nanoscale Reactors and Networks." *Annu. Rev. Phys. Chem.* Vol. 55, 2004, pp. 613–49.

[2] Krishnan, M., Namasivayam, V., Lin, R., Pal, R., Burns, M. A., "Microfabricated Reaction and Separation Systems." *Current Opinions in Biotechnology*, Vol. 12, 1, 2001, 92-98.

[3] Nguyen, T. Q., Wu, J., Doan, V., Schwartz, B. J., Tolbert, S. H., "Control of Energy Transfer in Oriented Conjugated Polymer Mesoporous Silica Composites." *Science*, Vol. 288, 2000, pp.652-656.

[4] Weng, G. Z.; Bhalla, U. S; Iyengar, R., "Complexity in Biological Signaling Systems." *Science*, Vol. 284,1999, pp. 92–96.

[5] Thomas, J. K., Ellison, E. H., "Various Aspects of the Constrains Imposed on the Photochemistry of Systems in Porous Silica." *Advances in Colloid and Interface Science*, Vol. 89-90, 2001, pp.195-238.

[6] Knudsen, M., *Kinetic Theory of Gases-Some Modern Aspects, Methuen's Monographs on Physical Subjects*, London, 1952.

[7] Ruetten, S. A., Thomas, J. K., "Fluorescence and Triplet Quantum Yields of Arenes on Surfaces." *J. Phys. Chem. B.,* Vol. 102, 1998, pp. 598-606.

[8] Fendler, J. H., *Membrane Mimitic Chemistry*, Wiley, New York, 1982.

[9] Antonietti, M., Landfester, K., Mastai, Y., "The Vision of 'Nanochemistry', or Is There a Promise for Specific Chemical Reactions in Nano-Restricted Environments?" *Israel Journal of Chemistry*, Vol. 41, 1, 2001, pp.1-5.

[10] Shtykov, S. N. "Chemical Analysis in Nanoreactors: Main Concepts and Applications." *Journal of Analytical Chemistry* (Translation of Zhurnal Analiticheskoi Khimii), Vol. 57(10), 2002, pp. 859-868.

[11] Li, Y., Zhang, M., Zhao, B., Zhang, S., Yang, M., "Advances in Nanoreactors." *Gaofenzi Tongbao*, Vol. 1, 2002, pp. 24-33.

[12] Fendler, J., "Self Assembled Nanostructured Materials." *Chemistry of Materials*, Vol. 8, 8, 1996, pp. 1616-1624.

[13] Vriezema, D., Aragones, M., Elemans, J., Cornlissen, J., Rowan, A., Nolte, R., "Self-Assembled Nanoreactors," *Chemical Reviews*, Vol. 105, 4, 2005, pp. 1445-1489.

[14] Sauer, M., Meier, W., "Colloidal Nanoreactors and Nanocontainers." *Colloids and Colloid Assemblies*, 2004, pp. 150-174.

[15] Volkmer, D., "From Biominerals to Biomimetic Materials." *Initiativen zum Umweltschutz*, Vol. 41, 2002, pp. 107-118.

[16] Ivanchev, S., Ozerin, A., "Vysokomolekulyarnya Soedineniya." *Seriya, A I Serya B*, Vol. 48, 8, 2006, pp. 1531-1544.

[17] Khomutov, G., "Interfacially Formed Organized Planar, Polymeric and Composite Nanostructures." *Advances in Colloid and Interface Science*, Vol. 1-21, 2004, pp. 79-116.

[18] Nardin, C., Meier, W., "Polymerizable Amphiphilic Block Copolymers: From Nanostructured Hydrogels to Nanoreactors and Ultrathin Films." *Chimia*, Vol. 55, 3, 2001, pp. 142-146.

[19] Mohwald, H., Lichtenfeld, H., Moya, S., Voigt, A., Baumler, H., Sukhorov, G., Caruso, F., Donath, E., "From Polymeric Films to Nanoreactors." *Macromolecular Symposia Proceeeding*, 1999, 145(Polymer Sorption Phenomena), 75-81.

[20] Zakharova, L., Semenov, V., Voronin, M., Valeeva, F., Ibragimova, A., Giniatullin, R., Chernova, A., Serey, V., Kudryavtseva, L., Latypov, S., Reznik, V., Konovalov, A., "Nanoreactors Based on Amphiphilic Uracilophans: Self-Organization and Reactivity Study." *Journal of Physical Chemistry B*, Vol. 111, 51, 2007, pp. 14152-14162.

[21] Ryu, E., Cho, H., Zhao, Y., "Catalyzing Methanolysis of Alkyl Halides in the Interior of an Amphiphilic Molecular Basket." *Organic Letters*, Vol. 9, 25, 2007, pp. 5147-5150.

[22] Ananchenko, G., Udachin, K., Ripmeester, J., Perrier, T., Coleman, A., "Phototransformation of Stilbene in van der Waals Nanocapsules." *Chemistry* (Weinheim an der Bergstrasse, Germany), Vol. 12,9, 2006, pp. 2441-7.

[23] Nishioka, Y., Yamaguchi, T., Yoshizawa, M., Fujita, M., "Unusual [2+4] and [2+2] Qycloadditions of Arenes in the Confined Cavity of Self-Assembled Cages." *Journal of the American Chemical Society*, Vol. 129, 22, 2007, pp. 7000-7001.

[24] Liu, Y., Liu, X., Warmuth, R., "Multicomponent Dynamic Covalent Assembly of a Rhombicubooctahedral Nanocapsule." *Chemistry: A European Journal*, Vol. 13, 32, 2007, pp. 8953-8959.

[25] Avram, L., Cohen, Y., "Molecules at Close Range: Encapsulated Solvent Molecules in Pyrogallol[4]arene Hexameric Capsules." *Organic Letters*, Vol. 8, 2, 2006, pp. 219-222.

[26] Daehne, L., Leporatti, S., Donath, E., Moehwald, H., "Fabrication of Micro Reaction Cages with Tailored Properties." *Journal of the American Chemical Society*, Vol. 123, 23, 2001, pp. 5431-5436.

[27] Lvov, Y., Caruso, F., "Biocolloids with Ordered Urease Multilayer Shells as Enzymatic Reactors." *Analytical Chemistry*, Vol. 73, 17, 2001, pp. 4212-4217.

[28] Neumann, T., Haupt, B., Ballauff, M., "High Activity of Enzymes Immobilized in Colloidal Nanoreactors." *Macromolecular Bioscience*, Vol. 4, 1, 2004, pp. 13-16.

[29] Jang, J., Oh, J., "Facile Fabrication of Photochromic Dye/Conducting Polymer Core/Shell Nanomaterials and their Photoluminescence." *PMSE Preprints*, Vol. 89, 2003, pp. 397-398.

[30] Labiguerie, J., Gredin, P., Mortier, M., Patriarche, G., de Kozak, A., "Synthesis of Fluoride Nanoparticles in Non-Aqueous Nanoreactors. Luminescence Study of Eu3+:CaF2. "*Zeitschrift fuer Anorganische und Allgemeine Chemie*, Vol. 632, 8-9, 2006, pp.1538-1543.

[31] Zhang, H., Piacham, T., Drew, M., Patek, M., Mosbach, K., Ye, L., "Molecularly Imprinted Nanoreactors for Regioselective Huisgen 1,3-Dipolar Cycloaddition Reaction." *Journal of the American Chemical Society*, Vol. 128,13, 2006, pp. 4178-4179.

[32] Balaji, R., Boileau, S., Guerin, P., Grande, D., "Porous Materials Derived from Poly (Caprolactone)-Poly(Methyl Methacrylate) Based Interpenetrating Polymer Networks by Selective Aminolysis and Photolysis." *Polymer Preprints*, Vol. 46, 1, 2005, pp. 300-301.

[33] Goltner, C.. "Inorganic Nanostructure Design with Amphiphilic Block Copolymers." *Surfactant Science Series*, 2001, pp. 797-818.

[34] Bronstein, L., Sidorov, S., Valetsky, P., "Nanostructured Polymeric Systems as Nanoreactors for Nanoparticle Formation." *Russian Chemical Reviews*, Vol. 73, 5, 2004, pp. 501-515.

[35] Ciebien, J., Clay, R., Sohn, B., Cohen, R., "Brief Review of Metal Nanoclusters in Block Copolymer Films," *New Journal of Chemistry*, Vol. 22, 7, 1998, pp. 685-691.

[36] Hamley, I. "Nanostructure Fabrication Using Block Copolymers." *Nanotechnology*, Vol. 14, 10, 2003, pp. R39-R54.

[37] Jerome, C. "Polymers for the Synthesis and Construction of Nanomaterials." *Chimie Nouvele*, Vol. 23, 90, 2005, pp. 119-124.

[38] Kita-Tokarczyk, K., Grumelard, J., Haefele, T., Meier, W., "Block Copolymer Cesicles Using Concepts from Polymer Chemistry to Mimic Biomembranes."Vol. 46, 11, 2005, pp.3540-3563.

[39] Yue, J., Cohen, R., "Nanoreactors for Inorganic Cluster Synthesis." *Supramolecular Science*, Vol. 1, 2, 1994, pp. 117-22.

[40] Kane, R., Cohen, R., Silbey, R., "Synthesis of PbS Nanoclusters within Block Copolymer Nanoreactors." *Chemistry of Materials*, Vol. 8,8, 1996, pp. 1919-1924.

[41] Fujita, Y., Ueno, K., Satomi, T., Yajima, H., Otsuka, H., "Physicochemical Characterization of the Py-g-PEG Copolymer at the Interface." *Transactions of the Materials Research Society of Japan*, Vol. 31, 3, 2006, pp. 649-653.

[42] Wang, Y., Wei, G., Zhang, W., Jiang, X., Zheng, P., Shi, L., Dong, A. "Responsive Catalysis of Thermoresponsive Micelle-Support Gold Nanoparticles." *Journal of Molecular Catalysis, A: Chemical*, Vol. 266, 1-2, 2007, pp. 233-238.

[43] Cong, Y., Fu, J., Zhang, Z., Cheng, Z., Xing, R., Li, J., Han, Y., "Fabrication of Arrays of Silver Nanoparticle Aggregates by Microcontact Printing and Block Copolymer Nanoreactors." *Journal of Applied Polymer Science*, Vol. 100, 4, 2006, pp. 2737-2743.

[44] Zhao, H., Douglas, E., "Salt-Induced Block Copolymer Micelles as Nanoreactors for the Formation of CdS Nanoparticles." *Materials Research Society Symposium Proceedings 2002*, pp. 43-48.

[45] Kane, R., Cohen, R., Silbey, R. "Synthesis of Doped ZnS Nanoclusters within Block Copolymer Nanoreactors." *Chemistry of Materials*, Vol. 11, 1, 1999, 90-93.

[46] Kane, R., Cohen, R., Silbey, R., Kuno, M., Bawendi, M., "Photoluminescent Mn-Doped ZnS Nanoclusters Synthesized within Block Copolymer Nanoreactors." *Materials Research Society Symposium Proceedings (1997)*, pp. 313-317.

[47] Manziek, L., Langenmayr, E., Lamola, A., Gallagher, M., Brese, N., Annan, N., "Functionalized Emulsion and Suspension Polymer Particles: Nanoreactors for the Synthesis of Inorganic Materials." *Chemistry of Materials*, Vol. 10, 10, 1998, pp. 3101-3108.

[48] Liu, S., Weaver, J., Save, M., Armes, S., "Synthesis of pH-Responsive Shell Cross-Linked Micelles and Their Use as Nanoreactors for the Preparation of Gold Nanoparticles." *Langmuir*, Vol. 2002, 18, 22, pp 8350-8357.

[49] Cheng, F., Yang, X., Peng, H., Chen, D., Jiang, M., "Well Controlled Formation of Polymeric Micelles with a Nanosized Aqueous Core and their Applications as Nanoreactors." *Macromolecules*, Vol. 40, 22, 2007, pp. 8007-8014.

[50] Nardin, C., Widmer, J., Winterhalter, M., Meier, W., "Amphiphilic Block Copolymer Nanocontainers as Bioreactors." *European Physical Journal E: Soft Matter*, Vol. 4, 4, 2001, pp. 403-410.

[51] Cresce, A., Silverstein, J., Bentley, W., Kofinas, P., "Nanopatterning of Recombinant Proteins Using Block Copolymer Templates." *Macromolecules*, Vol. 39, 17, 2006, pp. 5826-5819.

[52] Ge, Z., Xie, D., Chen, D., Joang, X., Zhang, Y., Liu, S., "Stimuli-Responsive Double Hydrophilic Block Copolymer Micelles with Switchable Catalytic Activity." *Macromolecules*, Vol. 40, 10, 2007, pp. 3538-3546.

[53] Zakharova, L., Ibragimova, A., Valeeva, F., Zakharov, A., Mustafina, A., Kudryavtseva, L., Harlampidi, H., Konovalov, A., "Nanosized Reactors Based on Polyethyleneimines: From Microheterogenous Systems to Immobilized Catalysts." *Langmuir*, Vol. 23, 6, 2007, pp. 3214-3224.

[54] Ranquin, A., Versees, W., Meier, W., Steyaert, J., Van Gelder, P., "Therapeutic Nanoreactors: Combining Chemistry and Biology in a Novel Triblock Copolymer Drug Delivery System." *Nano Letters*, Vol.5, 11, 2005, pp. 2220-2224.

[55] Nardin, C., Thoeni, S., Widmer, J., Winterhalter, M., Meier, W., "Nanoreactors Based on (Polymerized) ABA-Triblock Copolymer Vesicles." *Chemical Communications*, Vol. 15, 2000, pp. 1433-1434.

[56] Broz, P., Driamov, S., Ziegler, J., Ben-Haim, N., Marsch, S., Meier, W., Hunziker, P., "Toward Intelligent Nanosize Bioreactors: A pH-Switchable, Channel-Equipped, Functional Polymer Nanocontainer." *Nano Letters*, Vol. 6, 10, 2006, pp. 2349-2353.

[57] Horiuchi, S., Fujita, T., Hayakawa, T., Nakao, Y., "Three-Dimensional Nanoscale Alignment of Metal Nanoparticles Using Block Copolymer Films as Nanoreactors." *Langmuir*, Vol. 19, 7, 2003, pp. 2963-2973.

[58] Sohn, B., Seo, B., Yoo, S., "Changes of the Lamellar Period by Nanoparticles in the Nanoreactor Scheme of Thin Films of Symmetric Diblock Copolymers." *Journal of Materials Chemistry*, Vol. 12, 6, 2002, pp. 1730-1734.

[59] Li, L., Yokoyama, H., Nemeto, T., Sugiyama, K., "Facile Fabrication of Nanocellular Block Copolymer Thin Filmer Using Supercritical CO2." *Advanced Materials*, 2004, 16, 14, pp. 1226-1229.

[60] Boontongkong, Y. Cohen, R. E. "Cavitated Block Copolymer Micellar Thin Films: Lateral Arrays of Open Nanoreactors. Macromolecules," 2002, Vol. 35, pp. 3647-3652.

[61] Ho, R., Chen C., Chiang, Y., Ko, B., Lin, C., "Tubular Nanostructures from Degradable Core-Shell Cylinder Microstructures in Chiral Diblock Copolymers." *Advanced Materials*, Vol. 18, 18, 2006, pp. 355-2358.

[62] Cohen, R., "Block Copolymer Based Strategies for Controlling the Arrangement of Open and Closed Nanoreactors on Planar Substrates." *Polymer Preprints*, Vol. 48, 1, 2007, p. 869.

[63] Kaestle, G., Boyen, H., Weigl, F., Lengl, G., Herzog, T., Ziemann, P., Riethmueller, S., Mayer, O., Hartmann, C., Spatz, J., Moeller, M., Ozawa, M., Banhart, F., Garnier, M., Oelhafen, P., "Micellar Nanoreactors-Preparation and Characterization of Hexagonally Ordered Arrays of Metallic Nanodots." *Advanced Functional Materials*, Vol. 13, 11, 2003, pp. 853-861.

[64] Bronstein, L., Kraemer, E., Berton, B., Burger, C., Foerster, S., Antonietti, M., "Successive Use of Amphiphilic Block Copolymers as Nanoreactors and Templates: Preparation of Porous Silica with Metal Nanoparticles." *Chemistry of Materials*, Vol. 11, 6, 1999, pp. 1402-1405.

[65] Vamvakaki M., Papoutsakis L., Katsamanis V., Afchoudia T., Fragouli P., Iatrou H., Hadjichristidis N., Armes S., Sidorov S., Zhirov D., Zhirov V., Kostylev M., Bronstein L., Anastasiadis S., "Micellization in pH-Sensitive Amphiphilic Block Copolymers in Aqueous Media and the Formation of Metal Nanoparticles." *Faraday Discussions*, Vol. 128, 2005, pp. 129-47.

[66] Bronstein, L., Sidorov, S., Valetsky, P., Hartmann, J., Coefen, H., Antonietti, M., "Induced Micellization by Interaction of Poly 2-Vinylpyridine Block Poly Ethylene Oxide with Metal Compounds. Micelle Characteristics and Metal Nanoparticles Formation." *Langmuir*, 1999, 15, 19, 6256-6262.

[67] Wang, T., Rubner, M., Cohen, R., "Polyelectrolyte Multilayer Nanoreactors for Preparing Silver Nanoparticle Composites: Controlling Metal Concentration and Nanoparticle Size." *Langmuir*, Vol. 18, 8, 2002, pp. 3370-3375.

[68] Joly, S., Kane, R., Radzilowski, L., Wang, T., Wu, A., Cohen, R., Thomas, E., Rubner, M., "Multilayer Nanoreactors for Metallic and Semiconducting Particles." *Langmuir*, Vol. 16, 3, 2000, pp. 1354-1359.

[69] Choi W., Koo H., Park J., Kim D., "Synthesis of Two Types of Nanoparticles in Polyelectrolyte Capsule Nanoreactors and Their Dual Functionality." *Journal of the American Chemical Society*, Vol. 127, 46, 2005, pp. 16136-42.

[70] Vriezema, D., Garcia, P., Oltra, N., Natzakis, N., Kuiper, S., Nolte, R., Rowan, A., van Hest, J., "Positional Assembly of Enzymes in Polymerosome Nanoreactors for Cascade Reactions." *Angewandte Chemie, International Edition*, Vol. 46, 39, 2007, pp. 7378-7382.

[71] Nailani, M., de Hoog, H., Cornelissen, J., Palmans, A., van Hest, J., Nolte, R., "Polymerosome Nanoreactors for Enzymatic Ring Opening Polymerization." *Biomacromolecules*, Vol. 8, 12, 2007, pp. 3723-3728.

[72] Van Hest, J., Vriezema, D., Garcia, P., Cornelissen, J., Rowan, A., Nolter, R., "Enzyme Positional Assembly in Polymeric Capsules." *Polymer Preprints*, Vol. 47, 2, 2006, pp. 238.

[73] Rubner, M. F. "pH-Controlled Fabrication of Polyelectrolyte Multilayers: Assembly and Applications." *Multilayer Thin Films*, 2003, pp. 133-154.

[74] Shi, X., Shen, M., Moehwald, H., "Polyelectrolyte Multilayer Nanoreactors Toward the Synthesis of Diverse Nanostructured Materials." *Progress in Polymer Science*, Vol. 29, 10, 2004, pp. 987-1019.

[75] Sharma, G., Ballauff, M., "Cationic Spherical Polyelectrolyte Brushes as Nanoreactors for the Generation of Gold Particles." *Macromolecular Rapid Communications*, Vol. 25, 4, 2004, pp. 547-552.

[76] Zhang, M., Liu, L., Wu, C., Fu, G., Zhao, H., He, B., "Synthesis, Characterization and Application of Well-Defined Environmentally Responsive Polymer Brushes on the Surface of Colloid Particles." *Polymer*, Vol. 48, 7, 2007, pp. 1989-1997.

[77] Voegel, J., Decher, G., Schaaf, P., "Polyelectrolyte Multilayer Films in the Biotechnology Field." *Actualitie Chimique*, Vol. 11-12, 2003, pp. 30-38.

[78] Campas, M., Katakis, I., "Layer-by-Layer Nanostructures for the Construction of Bioelectrocatalytic Devices." *Trends in Electrochemistry and Corrosion at the Beginning of the 21st Century*, 2004, pp. 535-552.

[79] Zhang, J., Yeuming, Z., Zhiyuan, G., Zhishen, L., Liu, S., "Polyion Complex Micelles Possessing Thermoresponsive Coronas and Their Covalent Core Stabilization via Click Chemistry." *Macromolecules*, Vol. 41, 4, 2008, pp. 1444-1454.

[80] Chen, H., Zeng, G., Wang, Z., Zhang, X., Peng, M.,Wu, L., Tung, C., "To Combine Precursor Assembly and Layer-by-Layer Deposition for Incorporation of Single-Charged Species: Nanocontainers with Charge-Selectivity and Nanoreactors." *Chemistry of Materials*, Vol. 17, 26, 2005, pp. 6679-6685.

[81] Voronov A., Kohut A., Peukert W., "Synthesis of Amphiphilic Silver Nanoparticles in Nanoreactors from Invertible Polyester." *Langmuir*, Vol. 23, 1, 2007, pp. 360-3.

[82] Carrot, G., Valmalette, J., Plummer, C., Scholz, S., Dutta, J., Hofmann, H., Hilborn, J., "Gold Nanoparticle Synthesis in Graft Copolymer Micelles." *Colloid and Polymer Science*, Vol. 276, 10, 1998, pp. 853-859.

[83] Kohut, A., Voronov, A., Samaryk, V., Peukert, W., "Amphiphilic Invertible Polyesters as Reducing and Stabilizing Agents in the Formation of Metal Nanoparticles." *Macromolecular Rapid Communications*, Vol. 28, 13, 2007, pp. 1410-1414.

[84] Rossbach, B., Leopold, K., Weberskirch, R., "Self-Assembled Nanoreactors as Highly Active Catalysts in the Hydrolytic Kinetic Resolytion (HKR) of Epoxides in Water." *Angewandte Chemie, International Edition*, Vol. 45, 8, 2006, pp. 1309, 1312.

[85] Germain, M., Grube, S., Carriere, V., Richard-Foy, H., Winterhalter, M., Fournier, D., "Composite Nanocapsules: Lipid Vesicles Covered with Several Layers of Crosslinked Polyelectrolytes Advanced Materials." Vol. 18, 21, 2006, pp. 2868-2871.

[86] Michel, M., Vautier, D., Voegel, J., Schaaf, P., Ball, V., "Layer by Layer Self-Assembled Polyelectrolyte Multilayers with Embedded Phospholipid Vesicles." *Langmuir*, Vol. 20, 12, 2004, pp. 4835-4839.

[87] Voegel, J., "Multilayered Polyelectrolyte Films: Nanoreacting System and Film Degradation Tuning." *PMSE Preprints*, 2005, 93, 274.

[88] Frechet, J., "Dendritic Macromolecules at the Interface of Nanoscience and Nanotechnology." *Macromolecular Symposia*, 2003, 201, pp. 11-22.

[89] Gitsov, I., Hamzik, J., Ryan, J., Simonyan, A., Nakas, J., Omori, S., Krastenov, A., Cohen, T., Tannenbaum, S., "Enzymatic Nanoreactors for Environmentally Benign Biotransformation." *Biomacromolecules*, Vol., 9, 3, 2008, 804-811.

[90] Lee, J., Kin, K., "Rotaxane Dendrimers." *Topics in Current Chemistry*, 2003, 228, pp.11-140.

[91] Zhang Y., Chen Y., Niu H., Gao M., "Formation of CdS Nanoparticle Necklaces with Functionalized Dendronized Polymers." *Small (Weinheim an der Bergstrasse, Germany)*, Vol. 2,11, 2006, 1314-9.

[92] Zhao, M., Sun, L., Crooks, R., "Preparation of Cu Nanoclusters within Dendrimer Templates." *Journal of the American Chemical Society*, Vol. 120, 19, 1998, pp. 4877-4878.

[93] Murata, M., Tanaka, Y., Mizukagi, T., Ebitani, K., Kaneda, K., "Palladium-Platinum Bimetallic Nanoparticles Catalysts Using Dendron Assembly for Selective Hydrogenation of Dienes and Their Application to Thermomorphic System." *Chemistry Letters*, Vol. 34, 2005, pp. 272-273.

[94] Ooe, M., Murata, M., Mizugaki, T., Ebitani, K., Kaneda, K., *Nanoletters*, Vol. 2, 9, 2002, pp. 999-1002.

[95] Crespilho, F., Huguenin, F., Zucolotto, V., Olivi, P., Nart, F., Oliveira, O., "Dendrimers as Nanoreactors to Produce Platinum Nanoparticles Embedded in Layer-by-Layer Films for Methanol-Tolerant Cathodes." *Electrochemistry Communications*, Vol. 8, 2, 2006, pp. 348-352.

[96] Lu, Y., Mei, Y., Walker, R., Ballauff, M., Drechsler, M., "Nano-Tree-Type Spherical Polymer Brush Particles as Templates for Metallic Nanoparticles." *Polymer*, Vol. 47, 14, 2006, pp. 4985-4995.

[97] Liu, Z., Wang, X., Wu, H., Li, C., "Silver Nanocomposite Layer-by-Layer Films Based on Assembled Polyelectrolyte/Dendrimer." *Journal of Colloid and Interface Science*, Vol. 287, 2, 2005, pp. 604-611.

[98] Ornelas, C., Ruiz Aranzaez, J., Cloutet, E., Alves, S., Astruc, D., "Supramoleuclear Metallodendrimers as Nanoreactors for Molecular Recognition." *Angewandte Chemie, International Edition*, Vol. 46, 2007, pp. 872-877.

[99] Mizugaki, T., Miyauchi, Y., Murata, M., Ebitani, K., Kaneda, K., "Dendritic Nanoreactor Encapsulating Rh Complex Catalyst for Hyroformylation." *Chemistry Letters*, 34, 3, 2005, pp. 286-287.

[100] Khopade, A., Caruso, F., "Investigation of the Factors Influencing the Formation of Denrimer/Polyanion Multilayer Films." *Langmuir*, Vol. 18, 20, 2002, pp. 7668-7676.

[101] Li, J., Shi, L., An, Y., Li, Y., Chen, X., Dong, H., "Reverse Micelles of Star-Block Copolymer as Nanoreactors for Preparation of Gold Nanoparticles." *Polymer*, Vol., 47, 26, 2006, pp. 8480-8487.

[102] Heise, A., Hedrick, J., Trollsas, M., Hilborn, J., Frank, C., Miller, R., "Star-Like Amphiphilic Block Copolymers-Models for Unimolecular Micelles and Nanoreactors." *Polymer Preprints*, Vol. 40,11999, pp. 451-453.

[103] Hillmyer, M., Lodge, T., Li, Z., "Synthesis and Aqueous Self-Assembly of ABC Miktoarm Star Terpolymers." *Polymer Preprints*, Vol. 47, 2, 2006, pp. 786.

[104] Kosenko, O., Samchenko, Y., Ulberg, Z., Lukash, L., Kozineta, G., "Hydrogel Nanoreactors with Immobilized Mesenchymal Stem Cells." *Namosistemi, Nanomateriali, Nanoteknologii*, Vol., 5, 1, 2007, pp. 173-181.

[105] Samchenko, Y., Pasmurtseva, N., Ul'berg, Z., "Hydrogel Nanoreactors for Medical Purposes." *Dopovidi Natsional'noi Akademii Nauk Ukraini*, Vol. 2, 2007, pp. 146-150.

[106] Samchenko, Y., Poltoratskaya, T., Ulberg, Z. "Nanoreactors of Medical Purpose on the Base of Hydrogel Microgranules." *Namosistemi, Nanomateriali, Nanoteknologii*, Vol., 5, 1, 2007, pp. 93-101.

[107] Murthy, P., Mohan, Y., Murali, V., Sreedhar, B., Raju, K., "First Successful Design of Semi-IPN Hydrogen Siler Nanocmposites: A Facile Approach for Antibacterial Application." *Journal of Colloid and Interface Science*, Vol. 318, 2, 2008, pp. 217-224.

[108] Jiang, X., Xiong, D., Yingli, Z., Peiwen, Z., Wangqing, S., "Thermoresponsive Hydrogel of Poly (Glycidyl Methacrylate-co-N Isopropylacrylimide) as a Nanoreactor of Gold Nanoparticles." *Journal of Polymer Science A: Polymer Chemistry*, Vol. 45, 13, 2007, pp. 2812-2819.

[109] Palioura, D., Armes, S., Anastasiadis, S., Vamvakaki, M. "Metal Nanocrystals Incorporated within pH-Responsive Microgel Particles." *Langmuir*, Vol. 23, 10, 2007, pp. 5761-5768.

[110] Mohan, Y., Lee, K., Premkumar, T., Geckeler, K., "Hydrogel Networks as Nanoreactors: A Novel Approach to Silver Nanoparticles for Antibacterial Applications." *Polymer*, Vol. 48,1, 2007, pp. 158-164.

[111] Wang, H., Holmberg, B., Yan, Y., "Synthesis of Template-Free Zeolite Nanocrystals by Using in Situ Thermoreversible Polymer Hydrogels." *Journal of the American Chemical Society*, Vol., 125, 33, 2003, pp. 9928-9929.

[112] Yang, X., Feng, Y., Tian, G., Du, Y., Ge, X., Di, Y., Zhang, Y., Sun, B., Xiao, F., "Design and Size Control of Uniform Zeolite Nanocrystals Synthesized in Adjustable Confined Voids Formed by Recyclable Monodisperse Polymer Spheres." *Angewandte Chemie, International Edition*, Vol. 44, 17, 2005, pp. 2568.

[113] Antonietti, M., Grohn, F., Hartmann, J., Bronstein, L., "Nonclassical Shapes of Noble-Metal Colloids by Synthesis in Microgel Nanoreactors." *Angewandte Chemie, International Edition*, Vol. 36, 19, 1997, pp. 2080-2083.

[114] Kuzmany, H., Pfieffer, R., Simon, F., "Interior of Single-Shell Carbon Nanotubes. From Peapod to the Nanoreactor." *Physik in Unserer Zeit*, Vol. 39, 1, 2008, pp. 21-28.

[115] Peng, H., Jain, M., Li, Q., Peterson, D., Zhu, Y., Jia, Q., "Vertically Aligned Pearl-Like Carbonnanotube Arrays for Finer Spinning." *Journal of the American Chemal Society*, Vol. 130, 4, 2008, pp. 1130-1131.

[116] Wang, X., Liu, Z., Hu, P., Liu, Y., Han, B., Zhu, D., "Nanofluids in Carbon Nanotubes Using Supercritical CO2: A First Towards Nanochemical Reactions." *Applied Physics*, Vol. 80, 3, 2005, pp. 637-639.

[117] Makarova, L., Shabanova, I., Terebova, N., Kodolov, V., Semakina, N., Udmurt, G., "X-Ray Electron Investigation of Carbon Metal-Containing Nanostructures Obtained from Polymers in Condensed Medium." *Izvestiya Rossiiskoi Akademii Nauk, Seriya Fizicheskaya*, Vol. 69, 4, 2005, pp. 573-575.

[118] Hu, J., Bando, Y., Zhan, J., Zhi, C., Golberg, D., "Carbon Nanotubes as Nanoreactors for the Fabrication of Single-Crystalline Mg3N2 Nanowires." *Nanoletters*, Vol., 6, 6, 2006, pp. z1136-1140.

[119] Han, W., Kohler-Redlich, P., Scheu, C., Ernst, F., Ruhle, M., Grobert, N., Terrones, M., Kroto, H., Walton, D., "Carbon Nanotubes as Nanoreactors for Boriding Iron Nanowires." *Advanced Materials*, Vol. 12, 18, 2000, pp.1356-1359.

[120] Cao, F., Zhong, K., Gao, A., Chen, C., Li, Q., Chen, Q., "Reducing Reaction of Fe3O4 in Nanoscopic Reactors of a-CNTs." *Journal of Physical Chemistry B*, Vol. 111, 7, 2007, pp. 1724-1728

[121] Lipanov, A., Kodolov, V., Khkhryakov, N., DidikA., Kodolova, V., Semakina, N., "Problems of the Development of Nanoreactors to Synthesize Metal Nanoparticles in Carbon Shell." *Nauchno-Obraz*, Vol. 2, 2005, pp. 58-63.

[122] Miyawaki, J., Yudasaka, M., Imai, H., Yorimitsu, H., Isobe, H., Nakamura, E., Iijima, S., "Synthesis of Ultrafine Gd2O3 Nanoparticles Inside Single Wall Carbon Nanohorns." *Journal of Physical Chemistry B*, 2006, 110, 11, 5179-5181.

[123] Huang, J. "In Situ Observation of Quasimelting of Diamond and Reversible Graphite-Diamond Phase Transformations." *Nanoletters*, Vol. 7, 8, 2007, pp. 2335-2340.

[124] Deneke, C., Jin-Phillipp N., Loa, I., Schmidt, O., "Radial Superlattices and Single Nanoreactors." *Applied Physics Letters*. Vol. 84, 22, 2004, pp. 4475-4477.

[125] Graesr, M., Pippel, E., Greiner, A., Wendorff, J., "Polymer Core-Shell Finers with Metal Nanoparticle as Nanoreactors for Catalysis." Macromolecules, Vol. 40, 17, 2007, pp. 6032-6039.

[126] Fahmi, A., Pietsch, T., Gindy, N., "Hierarchical Nanoporous Structures by Self-Assembled Hybrid Materials Based on Block Copolymers." *Macromolecular Rapid Communications*, Vol. 28, 24, 2007, pp. 230-2305.

[127] Shang L., Wang Y., Huang L., Dong S., "Preparation of DNA-Silver Nanohybrids in Multilayer Nanoreactors by In Situ Electrochemical Reduction, Characterization, and Application." *Langmuir*, Vol. 23, 14, 2007, pp. 7738-44.

[128] Djalali R., Samson J., Matsui H., "Doughnut Shaped Nanoassemblies and Their Applications as Nanoreactors." *Journal of the American Chemical Society*, Vol. 126, 25, 2004, pp. 7935-9.

[129] Matsui, H., "Development of Nanoreactors via Peptide Self-Assemblies and Their Potential in Controlled Release and Enzymatic Catalysis Applications." *PMSE Preprints*, Vol. 90, 2004, p. 310.

[130] Sergeev-Cherenkov, A., Antipina, M., Yurova, T., Rakhnyanskaya, A., Gainutdinov, R., Tolstikhina, A., Kislov, V., Khomutov, G., "Low-Dimensional Hybrid Organic-Inorganic Nanostructures via Planar DNA-Amphiphilic Polycation Complexes." *Surface Science*, Vol. 566-568, 2004, 659-663.

[131] Antipina, M., Gainutdinov, R., Rakhnyanskaya, A., Sergeev-Cerenkov, A., Tolstikhina, A., Yurowa, T., Kislov, V., Khomutov, G., "DNA Complexed at the Surface of the Aqueous Phase: Novel Planar Polymeric and Composite Nanostructures." Biofizika, Vol. 48, 6, 2003, pp. 998-1010.

[132] Khomutov, G., Antipina, M., Gainutdinov, R., Kislov, V., Rakhnyanskaja, A., Segeev-Cherenkov, A., Tolstikhina, A., Yurova, T., "Interfacially Organized DNA/Polycation Complexes: A Route to New Planar Polymeric and Composite Nanostructures." *Materials Science and Engineering*, Vol. C23, 6-8, 2003, pp. 903-908.

[133] Shin, S., Steffenson, M., Claridge, T., Bayley, H., "Formation of a Chiral Center and Pyramidal Inversion at the Single Molecule Level." *Angewandte Chemie, International Edition*, Vol., 46, 39, 2007, pp. 7412-7416.

[134] Fang, M., Grant, P., McShane, M., Sukhorov, G., Golub, V., Lvov, Y., "Magnetic Bio/Nanoreactor with Multilayer Shells of Glucose Oxidase and Inorganic Nanoparticles." *Langmuir*, Vol. 18, 16, 2002, pp. 6338-6344.

[135] Fierobe, H., Mechaly, A., Tardif, C., Belaich, A., Lamed, R., Shoham, Y., Belaich, J., Bayer, E., "Designer Nanosomes: Selective Engineering of Dockerin-Containing Enzymes into Chimeric Scaffoldins to Form Defined Nanoreactors." *Special Publication-Royal Society of Chemistry*, 2002, pp. 113-123.

[136] Allen, R., Nielson, R., Wise, D., Shear, J., "Catalytic Three Dimensional Protein Architectures." *Analytical Chemistry*, 77, 16, 2005, pp. 5089-5095.

[137] Loudwig, S., Bayley, H., "Photoisomerization of an Individual Azobenzene Molecule in Water: An On/Off Switch Triggered by Light at a Fixed Wavelength." *Journal of the American Chemical Society*, Vol. 128, 38, 2006, 12404-12405.

[138] He, J., Kunitake, T., Nakao, A., "Facile in Situ Synthesis of Noble Metal Nanoparticles in Porous Cellulose Fibers." *Chemistry of Materials*, Vol. 15, 23, 2003, pp. 4401-4406.

[139] Ariga, K., Lvov, Y., "Self Assembly of Functional Protein Multilayers: From Planar Films to Microtemplate Encapsulation." *Surfactant Science Series*, Vol. 110, 2003, pp. 367-391.

[140] Lee, S., Gao, X., Matsui, H., "Material Synthesis: Growth of Ga2O3 Nanoparticles on Peptide Assemblies as Nanoreactors." *Journal of the American Chemical Society*, Vol. 129, 10, 2007, pp. 2954-2958.

[141] Solans, C., Izquierdo, P., Nolla, J., Azemar, N., Garcia-Celma, M., "Nano-Emulsions." *Current Opinion in Colloid & Interface Science*, Vol. 10, 3/4, 2005, pp. 102-110.

[142] Landfester, K., "Designing Particles: Miniemulsion Technology and its Application in Functional Coating Systems." *European Coatings Journal*, Vol. 20-22, 2005, 12, pp. 24-25.

[143] Pileni, M., "Nanomaterials Production by Soft Chemistry." *Nanostructured Materials*, 2002, pp. 1-21.

[144] Chernova, R., Doronin, S., Myznikova, I., " Nanoreactor (Surfactant Micelles) Influence on Aniline Derivative Protonation in Condensation Reactions with Aldehydes."

Izvestiya Vysshikh Uchebnykh Zavedenii, Khimiya i Khimicheskaya Tekhnologiya, Vol. 48, 6, 2005, pp. 113-116.

[145] Liu, W., Kumar, J., Tripathy, S., Samuelson, L., "Enzymatic Synthesis of Conducting Polyaniline in Micelle Solutions." *Langmuir*, Vol. 18, 25, 2002, pp. 9696-9704.

[146] Samuelson, L., Nagarajan, R., Kumar, J., Bruno, F., Cholili, A, Tripathy, S., "Nanoreactors for the Enzymatic Synthesis of Conducting Polyaniline." *Synthetic Metals*, Vol. 19, 1-3, 2001, pp. 271-272.

[147] Kotsabasakis, V., Georgopoulou, E., Pitsikalis, M.., Hadjichristidis, N., Papadogianakis, G., "Catalytic Conversions in Aqueous Media: A Novel and Efficient Hydrogenation of Polybutadiene-1,4-Block-Poly(Ethylene Oxide) Catalyzed by Rh/TPPTS Complexes in Mixed Micellar Nanoreactors." *Journal of Molecular Catalysis*,1-2, 2005, 93-101, pp. 231.

[148] Martinez-Junza, V., Rizzi, A., Jolliffe, K., Head, N., Paddon-Row, M., Braslavsky, A., "Conformational and Photophysical Studies on Porphyrin-Containing Donor-Bridge-Acceptor Compounds. Charge Separation in Micellar Nanoreactors." *Physical Chemistry Chemical Physics* , Vol. 7, 24, 2005, pp. 4114-4125.

[149] Kobylyansay, E., Ishchuk, Y., Altshuler, M., "Fundamentals of Nanotechnologies of Lubricating Materials." *Kataliz I Neftekhimiya*, Vol. 13, 2005, p. 108.

[150] Maitra, A., Ghosh, P., Ghosh, P. De, T., Sahoo, S., Gaur, U., "Long Circulating RES Evading Hydrophilic Nanoparticles." *Proceeding of the International Symposium on Controlled Release of Bioactive Materials*, 1998, 25th, pp. 168-169.

[151] Witak, E., Kochanowski, A., Pazdro, M., Bortel, E., "Microemulsions as the Source of Nanolatexes and Nanoreactors." *Polimery*, Vol. 51, 7,8, 2006, pp. 507-516.

[152] Eastoe, J., Stebbing, S., Dalton, J., Heenan, R., "Preparation of Colloidal Cobalt Using Reversed Micelles." *Colloids and Surfaces*, Vol. 119, 2/3, 1996, pp. 123-131.

[153] Lisiecki, I., Pileni, M., " Synthesis of Well-Defined and Low Size Distribution Cobalt Nanocrystals: The Limited Influence of Reverse Micelles." *Langmuir*, Vol. 19, 22, 2003, pp. 9486-9489.

[154] Carpenter, E., O'Connor, C., Harris, V., "Atomic Structure and Magnetic Properties of MnFe2O4 Nanoparticles Produced by Reverse Micelle Synthesis." *Journal of Applied Physics*, Vol. 85(8, Pt. 2A), 1999, pp. 5175-5177.

[155] Ingert, D., Pileni, M., "Limitations in Producing Nanocrystals Using Reverse Micelles as Nanoreactors." *Advanced Functional Materials*, Vol. 11, 2, 2001, pp. 136-139.

[156] Taguchi, M., Yagi, I., Nakagawa, M., Iyoda, T., Einaga, Y., "Photocontrolled Magnetization of CdS-Modified Prussian Blue Nanoparticles." *Journal of the American Chemical Society*, Vol. 128, 33, 2006, pp. 10978-10982.

[157] Chiang, C., Hsu, M., Lai, L., "Control of Nucleation and Growth of Gold Nanoparticles in AOT/Span80/Isooctane Mixed Reverse Micelles." *Journal of Solid State Chemistry*, Vol. 177, 11, 2004, pp. 3891-3895.

[158] Vaidya, S., Ahmad, T., Agarwal, S., Ganguli, A., "Nanocrystalline Oxalate/Carbonate Precursors of Ce and Zr and Their Decompositions to CeO2 and ZrO2 Nanoparticles." *Journal of the American Ceramic Society*, Vol.90, 3, 2007, pp. 863-869.

[159] Castagnola, M., Dutta, P., "Synthesis of Microporous Faujasitic-Like Zincophosphates from Reverse Micelles." *Microporous and Mesoporous Materials*, Vol. 34, 1, 2000, pp. 61-65.

[160] Chakroborty, S., Sahoo, B., Teraoka, I., Gross, R., *Polymer Preprints*, 2003, 44,2, pp. 602-603.

[161] Sui, C., Chu, Y., Yu, M., Liu, C. "Self Organization of Spherical PANI/TiO2 Nanocomposites in Reverse Micelles." *Colloids and Surfaces*, Vol. 251, 103, 2004, pp. 103-107.

[162] Lee, Y., Lee, J., Bae, C., Park, J., Noh, H., Park, J., Hyeon, T., "Large-Scale Synthesis of Uniform and Crystalline Magnetite Nanoparticles Using Reverse Micelles as Nanoreactors under Reflux Conditions." *Advanced Functional Materials*, Vol. 15, 12, 2005, p. 2036.

[163] Duarte, E., Itri, R., Lima, E., Baptista, M., Berquo, T., Goya, G., "Large Magnetic Anisotropy in Ferrihydrite Nanoparticles Synthesized from Reverse Micelles." *Nanotechnology*, Vol. 17, 22, 2006, pp. 5549-5555.

[164] Ahmad, T., Ganguli, A., "Reverse Micellar Route to Nanocrystalline Titanates (SrTiO3, Sr2TiO4, and PbTiO3): Structural Aspects and Dielectric Properties." *Journal of the American Ceramic Society*, Vol. 89, 4, 2006, pp. 1326-1332.

[165] Isobe, T., "High Efficiency of Luminescence on Nanosized Phosphors." *Oyo Butsuri*, Vol. 70, 9, 2001, pp. 1087-1091.

[166] Pileni, M., "The Role of Soft Colloidal Templates in Controlling the Size and Shape of Inorganic Nanocrystals." *Nature Materials*, Vol. 2, 3, 2003, pp. 145-50.

[167] Uskokovic, V., Drofenik, M., "Reverse Micelles: Inert Nanoreactors or Physico-Chemically Active Guides of the Capped Reactions." *Advances in Colloid and Interface Science*, Vol. 133, 1, 2007, pp. 23-34.

[168] Liveri, V., "Calorimetric Investigations of Solutions of Reversed Micelles." *Surfactant Science Series*, Vol. 93, 2001, pp. 1-22.

[169] Kumar, K., Brooks, D., "Comparison of Hyperbranched and Linear Polyglycidol Unimolecular Reverse Micelles as Nanoreactors and Nanocapsules." *Macromolecular Rapid Communications*, Vol. 26, 3, 2005, 155-159.

[170] Gao, H., Jones, M., Tewari, P., Ranger, M., Leroux, J., "Star-Shaped Alkylated Poly(Glycerol Methacrylate) Reverse Micelles: Synthesis and Evaluation of their Solubilizing Properties in Dichloromethane." *Journal of Polymer Science*, Vol. 45,12, 2007, pp. 2425-2435.

[171] Szaloki, M., Scribanek, R., Hartmann, J., Hegedus, C., Borbely, J., "Preparation of Reactive Polymeric Nanoparticles." *PMSE Preprints*, Vol. 94, 2006, pp. 783-784.

[172] Kumar, K., Brooks, D., "Amphiphilic Hyperbranched Polyglycidol Unimolecular Reverse Micelles as Nanoreactors." *PMSE Preprints*, Vol. 91, 2004, pp. 6-7.

[173] Bolinger, P., Stamou, D., Vogel, H., "Integrated Nanoreactor Systems: Triggering the Release and Mixing of Compounds Inside Single Vesicles." *Journal of the American Chemical Society*, Vol. 126, 28, 2004, pp.8594-8595.

[174] Sangregorio, C., Wiemann, J., O'Connor, C., Rosenzweig, Z., "A New Method for the Synthesis of Magnetoliposomes." *Journal of Applied Physics*, Vol. 85(8, Pt. 2B), 1999, pp. 5699-5701.

[175] Walde, P., Ichikawa, S., "Enzymes Inside Lipid Vesicles: Preparation, Reactivity, and Application." *Biomolecular Engineering*, Vol. 18, 4, 2001, pp. 143-177.

[176] Bota A., Varga, Z., Goerigk, G., "Biological Systems as Nanoreactors: Anomalous Small-Angle Scattering Study of the CdS Nanoparticle Formation in Multilamellar Vesicles." *The Journal of Physical Chemistry*, Vol., 111, 8, 2007, pp. 1911-5.

[177] Huysmans, G., Ranquin, A., Wyns, L., Steyaert, J., Van Gelder, P., "Encapsulation of Therapeutic Nucleoside Hydrolase in Functionalized Nanocapsules." *Journal of Controlled Release*, Vol. 102, 1, 2005, pp. 171-179.

[178] Lindemann, M., Winterhalter, M., "Membrane Channels as a Tool to Control Nanoreactors." *Nanobiotechnology*, Vol. 153, 4, 2006, pp. 107-111.

[179] Graff, A., Winterhalter, M., Meier, W., "Nanoreactors from Polymer Stabilized Liposomes." *Langmuir*, Vol. 17, 3, 2001, pp. 919-923.

[180] Vasilevskaya, V., Aerov, A., Khokhlov, A. "Control of Reactions between Surfactant Reagents in Miniemulsions. Surface Nanoreactors." *Colloid and Polymer Science*, Vol. 284, 5, 2006, pp. 459-467.

[181] Gardner, A., Vasquez, V., Clifon, A., Graeve, O., "Molecular Dynamics Analysis of the AOT Water Isooctane system: Effect of Simulation Time, Initial Configuration, and Model Salts." *Fluid Phase Equilibria*, Vol. 262,1-2, 2007, 264-270.

[182] Vasilievskaya, V., Aerov, A., Khokhlov, A., "Catalytic Reactions of a Surface-Active Catalyst and a Surface Active Substrate in Emulsions: The Optimal Drop Size." *Doklady Physical Chemistry*, Vol. 398, 2, 2004, pp. 258-261.

[183] Liu, J., Raveendran, P., Shervani, Z., Ikushima, Y., Hakuta, Y., "Synthesis of Ag and AgI Quantum Dots in AOT-Stabilized Water-in-CO2 Microemulsions." *Chemistry—A European Journal*, Vol. 11, 6, 2005, pp. 1854-1860.

[184] He, J., Ma, T., Huang, Y., Yang, J., "Study on preparation of spherical zirconia powder in microemulsion system and its surface characteristics. *Key Engineering Materials*, Vol. 280-283, 2005, pp. 977-980.

[185] Ma, T., Huang, Y., Yang, J., He, J., "Preparation of ultrafine zirconia spherical powder by nano-reactor of microemulsion." *Wuji Cailiao Xuebao*, Vol. 18, 5, 2003, pp. 1107-1112.

[186] Ma, T., Huang, Y., Yang, J., He, J., "Preparation of spherical zirconia powder by nanoreactor of microemulsion and its densification behavior." *Xiyou Jinshu Cailiao Yu Gongcheng*, Vol. 33,11,2004, pp. 112-1131.

[187] Huang, Y., Ma, T., Yang, J., Zhang, L., He, J., Li, H., "Preparation of spherical ultrafine zirconia powder in microemulsion system and its dispersibility." *Ceramics International*, Vol. 30, 5, 2004, pp. 675-681.

[188] Karanikolos, G., Alexandridis, P., Itskos, G., Petrou, A., Mountziaris, T., "Synthesis and Size Control of Luminescent ZnSe Nanocrystals by a Microemulsion-Gas Contacting Technique." *Langmuir*, Vol., 20, 3, 2004, pp. 550-553.

[189] Kuo, T., Tung, C., Lin, Y., Jang, S., "Supercritical microemulsions as nanoreactors for manufacturing ZnS nanophosphors." *Chemistry Letters*, Vol. 33, 7, 2004, pp. 802-803.

[190] Calandra, P., Goffredi, M., Liveri, V., "Study of the growth of ZnS nanoparticles in water/AOT/n-heptane microemulsions by UV-absorption spectroscopy." *Colloids and Surfaces*, Vol. 160, 1, 1999, pp. 9-13.

[191] Liveri, V., Rossi, M., D'Arrigo, G., Manno, D., Micocci, G., "Synthesis and characterization of ZnS nanoparticles in water/AOT/n-heptane microemulsions." *Applied Physics*, Vol. 69, 4, 1999, pp. 369-373.

[192] Hingorani, S., Shah, D., Multani, M., "Effect of process variables on the grain growth and microstructure of ZnO-Bi2O3 varistors and their nanosize ZnO precursors." *Journal of Materials Research*, Vol. 10, 2, 1995, pp. 461-7.

[193] Kim, D., Oh, S., Lee, J., "Preparation of Ultrafine Monodispersed Indium-Tin Oxide Particles in AOT-Based Reverse Microemulsions as Nanoreactors." *Langmuir*, Vol. 15, 5, 1999, pp. 1599-1603.

[194] Ohde, H., Ohde, M., Bailey, F., Kim, H., Wai, C., Water-in-CO2 Microemulsions as Nanoreactors for Synthesizing CdS and ZnS Nanoparticles in Supercritical CO2." *Nano Letters*, Vol. 2, 7, 2002, pp. 721-724.

[195] Fernandez, C., Wai, C., "Continuous tuning of cadmium sulfide and zinc sulfide nanoparticle size in a water-in –super-critical CO2 microemulsion." *Chemistry: A European Journal*, Vol. 13, 20, 2007, pp. 5838-5844.

[196] Beck, C., Hartl, W., Hempelmann, R. "Size-controlled synthesis of nanocrystalline BaTiO3 by a sol-gel type hydrolysis in microemulsion-provided nanoreactors." *Journal of Materials Research*, Vol. 13, 11, 1998, pp. 3174-3180.

[197] Gross, B., Beck, C., Meyer, F., Krajewski, Th., Hempelmann, R., Altgeld, H., "BaZr0.85Me0.15O2.925 (Me=Y, In and Ga): Crystal growth, high-resolution transmission electron microscopy, high-temperature X-ray diffraction and neutron scattering experiments." *Solid State Ionics*, Vol. 145, 1-4, 2001, pp. 325-331.

[198] Lee, J., Lee, Y., Youn, J., Na, H., Yu, T., Kim, H., Lee, S., Loo, Y., Kwak, J., Park, H., Chang, H., Hwang, M., Park, J., Kim, J., Hyeon, T., "Simple synthesis of functionalized superparamagnetic magnetite/silica core/shell nanoparticle and their

application as magnetically separable high performance biocatalysts." *Small*, Vol. 4, 1, 2008, pp. 143-152.

[199] Crespy, D., Stark, M., Hoffmann-Richter, C., Ziener, U., Landfester, K., "Polymeric Nanoreactors for Hydrophilic Reagents Synthesized by Interfacial Polycondensation on Miniemulsion Droplets." *Macromolecules*, Vol. 40, 9, 2007, pp. 3122-3135.

[200] Sawant, P., Ramaniah L., "Capacity of nano-reactors of AOT micro-emulsions to form and sustain ultra small semiconductor quantum dots." *Journal of Nanoscience and Nanotechnology*, Vol., 6, 1, 2006, pp. 241-247.

[201] Manziek, L., Langenmayr, E., Lamola, A., Gallagher, M., Brese, N., Annan, N., "Functionalized Emulsion and Suspension Polymer Particles: Nanoreactors for the Synthesis of Inorganic Materials." *Chemistry of Materials*, Vol. 10, 10, 1998, pp. 3101-3108.

[202] Yaghmur, A., Aserin, A., Abbas, A., Garti, N., "Reactivity of furfural-cysteine model reaction in food-grade five-component nonionic O/W microemulsions." *Colloids and Surfaces*, Vol. 253, 1-3, 2005, pp. 223,234.

[203] Zakharova, L., Ibragimova, A., Valeeva, F., Kudryavtseva, L., Konovlov, A., Zakharov, A., Selivanova, N., Osipova, V., Strelkov, M., Glayametdinov, Y., "Self-organization and catalytic activity of the poly(ethylene glycol) (10) Monododecyl ether/poly(ethyleneinine/lanthanum nitrate system." *Journal of Physical Chemistry C*, Vol. 111, 37, 2007, pp. 13839-13845.

[204] Groger, H., May, O., Huesken, H., Georgeon, S., Drauz, K., Landfester, K., "Enantioselective enzymatic reactions in miniemulsions as efficient nanoreactors." *Angewandte Chemie, International Edition*, Vol. 45, 10, 2006, pp. 1645-1648.

[205] Musyanovych, A., Mailaender, V., Landfester, K., "Miniemulsion Droplets as Single Molecule Nanoreactors for Polymerase Chain Reaction." *Biomacromolecules*, Vol. 6, 4, 2005, pp. 1824-1828.

[206] Okhonin, V., Liu, X., Krylov, S., "Transverse diffusion of laminar flow profiles to produce capillary nanoreactors." *Analytical Chemistry*, Vol. 77, 18, 2005, pp. 5925-5929.

[207] Lee, S., Lee, J., Cheong, I., Lee, H., Kim, J., "A facile route of polythiophene nanoparticles via Fe3+ catalyzed oxidative polymerization in aqueous medium." *Journal of Polymer Science*, Vol. 46,6, 2008, pp. 2097-2109.

[208] Tokumoto, M., Surendran, G., dos Santos, E., Kooyman, P., Remita, H., Ramos, L., Prouzet, E., "Synthesis of nanomaterials in highly swollen liquid crystals." *Materials Research Society Symposium Proceedings*, Vol. Organic/Inorganic Hybrid Materials—2004, 2005, 2004, pp. 847107-118.

[209] Fang, J. "Ordered arrays of self-assembled lipid tubules: Fabrication and applications." *Journal of Materials Chemistry*, Vol. 17, 33, 2007, pp. 3479-3484.

[210] Pena dos Santos, E., Tokumito, M., Surendran, G., Remita, H., Bourgaux, C., Diedonne, P., Prouzet, E., Ramos, L., "Existence and stability of new nanoreactors:

Highly swollen hexagonal liquid crystals." *Langmuir*, Vol. 21,10, 2005, pp. 4362-4369.

[211] Castillo M., Fousse L., Fraile J., Garcia J., Mayoral J., "Supported ionic-liquid films (SILF) as two-dimensional nanoreactors for enantioselective reactions: surface-mediated selectivity modulation (SMSM)." *Chemistry*, Vol. 13,1, 2007, pp. 287-91.

[212] Dellinger, T., Braun, P., "Lyotropic liquid crystals as nanoreactors for nanoparticle synthesis." *Chemistry of Materials*, Vol. 16, 11, 2004, pp. 2201-2207.

[213] Dellinger, T., Braun, P., "BiOCl nanoparticles synthesized in lyotropic liquid crystal nanoreactors." *Scripta Materialia*, 44(8/9), 2001, 1893-1897.

[214] Surendran, G., Ramos, L., Pansu, B., Prouzet, E., Beaunier, P., Audonnet, F., Remita, H., "Synthesis of porous platinum nanoballs in soft templates." *Chemistry of Materials*, Vol. 19, 21, 2007, pp. 5045-5048.

[215] Surendran, G., Tokumoto, M., Dos Santos, E.,; Remita, H., Ramos, L., Kooyman, P., Santilli, C., Bourgaux, C., Dieudonne, P., Prouzet, E., "Nanomaterials. Institut Europeen des Membranes", *Chemistry of Materials*, Vol. 17, 6, 2005, pp. 1505-1514.

[216] Jin, L., Bae, J., Ryu, J.,; Lee, M., "Ordered nanostructures from the self-assembly of reactive coil-rod-coil molecules." *Angewandte Chemie, International Edition*, Vol., 45, 4, 2006, pp. 650-653.

[217] Bakeeva, R., Kudryavtsev, D., Zakharova, L., Kudryavtseva, L., Raevska, A., Sopin, V., "Micellar, liquid crystalline and polymer systems based on surfactant and polyethylene imine as nanoreactors for the transfer of phosphoryl group." *Molecular Crystals and Liquid Crystals*, Vol. 367, 2001, pp. 585-596.

[218] Ile, K., Schaaf, G., Bankaitis, V., "Phosphatidylinositol transfer proteins and cellular nanoreactors for lipid signaling." *Nature Chemical Biology*, Vol. 2, 11, 2006, pp. 576-583.

[219] Cornellas-Aragones, M., Englekamp, H., Clasessen, V., Sommerdjik, N., Rowan, A., Christianen, P., Maan, J., Verdui, B., Cornelissen, J., Nolte, R., "A virus based single enzyme nanoreactor." *Nature Nanotechnology*, Vol. 2, 10, 2007, pp. 635-639.

[220] De Menoval, B., Ayrault, P., Gnep, N., Guisenet, M., "N-butene skeletal isomerization over HFER zeolites: influence of Si/Al ratio and carbonaceous deposits." *Applied Xatalysis, A. General*, Vol. 304, 2006, 1-13.

[221] De Menorval, A., Gnep, N., Guisnet, M., "Mechanism of n-butene skeletal isomerization over HFER zeolites: a new proposal." *Journal of Catalysis*, Vol. 230, 1, 2005, pp.38-51.

[222] Farzaneh, F., Taghavi, J., Malakooti, R., Ghandi, M., "Immobilized vitamin B12 within nanoreactors of MCM-41 as a selective catalyst for oxidation of organic substrates. "*Journal of Molecular Catalysis A.*, Vol. 244,1-2, 2006, 252-257.

[223] Yu, C., Fan, J., Tian, B., Zhao. D., Stucky, G., High yield synthesis of periodic mesoporous silica rods and their replication to mesoporous carbon rods." *Advanced Materials*, Vol. 14, 23, 2002, pp. 1742-1745.

[224] Han, B., Polarz, S., Antonietti, M., "Cyclodextrin-based porous silica materials as in situ chemical "nanoreactors" for the preparation of variable metal-silica hybrids." Chemistry of Materials, Vol. 13, 11, 2001, pp. 3915-3919.

[225] Doshi, D., Huesing, N., Lu, M., Fan, H., Lu, Y., Simmons-Potter, K., Potter, B., Hurd, A., Brinker, C., "Optically define multifunctional pattering of photosensitive think film silica mesophases." *Science*, Vol. 290, 5489, 2000, pp. 107-111.

[226] Chauan, B., Latif, U., "Inorganic.organic hybrid nanoreactors based on cyclic and cubic siloxane scaffolds." *Macromolecules*, Vol. 38, 152005, pp. 6231-6235.

[227] Tsang, S., Zhang, N., Fellas, L., Steele, A., "Supported micelles in biphasic oxidation catalysis." *Catalysis Today*, Vol. 61,1-4, 2000, 29-36.

[228] Jungmann, N., Schmidt, M., Maskos, M., "Amphiphilic Poly(organosiloxane) Nanospheres as Nanoreactors for the Synthesis of Topologically Trapped Gold, Silver, and Palladium Colloids." *Macromolecules*, Vol. 36, 11, 2003, pp. 3974-3979.

[229] Kolesnik, I., Eliseev, A., Garshev, A., Lukashin, A., Tretyakov, Y., "Synthesis of silver nanoparticles in mesoporous high aluminum aluminosilicate matrices." *Russian Chemical Bulletin*, Vol. 53, 11, 2004, pp. 2496-2498.

[230] Mazali, I., Romano, R., Alves, O., "Integrated chemical systems built using nanoporous glass/ceramics as substrates." *Thin Solid Films*, Vol. 495, 1-2, 2006, pp. 64-67.

[231] Casas, L., Roig, A., Rodriguez, E., Molins, E., Tejada, J., Sort, J., "Silica aerogel-iron oxide nanocomposites :structural and magentic properties." *Journal of Nano-Crystalline Solids*, Vol. 285, 1-3, 2001, 37-43.

[232] Yang, X., Li, Z., Liu, B., Klein-Hofmann, A., Tian, G., Feng, Y., Ding, Y., Su, D., Xiao, F., "'Fish-in-net' encapsulation of enzymes in macroporous cages as stable, reusable, and active heterogeneous biocatalysts." *Advanced Materials*, 2006, 18, 4, 410-414.

[233] Malakooti, R., Farzaneh, F., Ghandi, M.,"Synthesis, characterization and studies on catalytic behavior of Mn(II) complex with 2.2' bipyridine 1,1'-dioxide ligand within nanoreactors of MCM-41." *Journal of Sciences, Islamic republic of Iran.* Vol. 17,1, 2006, 43-52.

[234] Lee, Y.-S., Yu, H., Kwon, O.-H., Jang, D.-J. "Excited-state deprotonation dynamics of 2-naphthol in NaX nanoreactors." *Studies in Surface Science and Catalysis*, Vol. 156, 2005, pp.741-746.

[235] Parvelescu, V., Constantin, C., Su, B., "Liquid phase oxidation of aromatic hydrocarbons using highly ordered nb and NbCo-MCM-41 nanoreactors." *Journal of Molecular Catalysis A:Chemical*, Vol. 202,1-2, 2003, pp. 171-178.

[236] Franco, M., Rosenbach, N., Ferrier, G., Guerra, A., Kover, W., Turci, C., Mota, C., "Rearrangement, nucleophilic substitution and halogen switch reactions of alkyl halides over NaY zeolite, formation of the bicyclobutonium cation inside the zeolite cavity." *Journal of the American Chemical Society*, Vol. 130, 5, 2008, pp. 1592-1600.

[237] Eliseev, A., Napolskii, K., Kolesnik, I., Kolenko, Y., Lukashin, A., Gornet, P., Tretyakov, Y., "Mesoporous aluminosilicates as a host and reactor for preparation of ordered metal nanowires." *NATO Science Series, II: Mathematic, Physics and Chemistry*, Vol.152, 2004, pp. 109-122.

[238] Urban, M., Konya, Z., Mehn, D., Zhu, J., Kiricsi, I., "Mesoporous silicates as nanoreactors for carbon nanotube production in the absence of transition metal catalysts." *Journal of Nanoscience and Nanotechnology*, Vol. 3, 1-2, 2003, pp. 111-119.

[239] Zu, C., Yu, W., Zhou, X., Zhao, D., Yang, P., "Rapid communication in mass spectrometry." Vol. 20, 20, 2006, pp. 3139-3144.

[240] Fan, J., Shui, W., Yang, P., Wang, X., Xu, Y., Wang, H., Chen, X., Zhao, D., *Chemistry: A European Journal*, Vol. 11,18,2005, pp. 5391-5396.

[241] Lin, H., Mou, C., "Structural and morphological control of cationic surfactant templated mesoporous silica." *Accounts of Chemical Research*, Vol. 35, 11, 2002, pp. 927-935.

[242] Miura K., Kawase, M., Ashida, R., Gerlach, I., Yamamoto, T., "Nano-reactor for producing high performance nanomaterials." *Chemical Engineering Science*, Vol. 62, 2007, 18-20.

[243] Vradman, L., Peer, Y., Mann-Kiperman, A., Landau, M.,"Thermal decomposition-precipitation inside the nanoreactors. High loading of W oxide nanoparticles into the nanotubes of SBA-15." *Studies in Surface Science and Catalysis*, Vol. 146 (Nanotechnology in Mesostructured Materials), 2003, pp. 121-124.

[244] Lee, C., Lin, T., Mou, C., "EPR studies of free radical reactions of C60 embedded in mesoporous MCM41 materials in aqueous solution." *Physical Chemistry Chemical Physics*, Vol. 4, 13, 2002, 3106-3111.

[245] Kidder, M., Buchanan, A., "Effect of pore confinement and molecular orientation on hydrogen transfer during a free radical reaction in mesoporous silica." *Journal of Physical Chemistry C*, Vol. 112,2008, 3027-3031.

[246] Pastore, H., Ozin, G., Poe, A., "Intrazeolite metal carbonyl kinetics:substitution reations of molybdenum carbonyl (Mo(12CO)6) in sodium zeolite Y." *Journal of the American Chemical Society*, Vol. 115,4,1993, 1215-1230.

[247] Zaker, Z. E., Farzaneh, F., Ghandi, M., Mohammadi, N., "Synthesis, characterization and epoxidation catalytic activity of vanadium sulfate immobilized mesoporous MCM-41." *Polish Journal of Chemistry*, Vol. 82, 3, 2008, pp. 613-619.

[248] Zuo, C., Yu, W., Zhou, X., Zhao, D., Yang, P., "Highly efficient enrichment and subsequent digestion of proteins in the mesoporous molecular sieve silicate SBA-15 for matrix-assisted laser desorption/ionization mass spectrometry with time-of-flight/time-of-flight analyzer peptide mapping." *Rapid Communications in Mass Spectrometry*, Vol. 20, 20, 2006, pp. 3139-3144.

[249] Shui, W., Fan, J., Yang, P.,; Liu, C., Zhai, J., Lei, J., Yan, Y., Zhao, D., Chen, X., "Nanopore-Based Proteolytic Reactor for Sensitive and Comprehensive Proteomic Analyses." *Analytical Chemistry*, Vol. 78, 14, 2006, pp. 4811-4819.

[250] Fan, J., Shui, W., Yang, P., Wang, X., Xu, Y., Wang, H., Chen, X., Zhao, D., "Mesoporous silica nanoreactors for highly efficient proteolysis." *Chemistry: A European Journal*, Vol. 11, 18, 2005, pp. 5391-5396.

[251] Nourouzi, F., Farzaneh, F., Khosrosheni, M., "Immobilized cobalt chloride within nanoreactors of Si-MCM-48 as selective catalysts for epoxidation of alkenes." *Reaction Kinetic and Catalysis Letters*, Vol. 89, 1, 2006, pp. 139-147.

[252] Tretyakov, Y., Lukashin, A., Eliseev, A., "Synthesis of functional nanocomposites based on solid-phase nanoreactors." *Russian Chemical Reviews*, Vol. 73, 9, 2004, pp. 819-821.

[253] Buso, D., Falcaro, P., Costacurta, S., Guglielmi, M., Martucci, A., Innocenzi, P., Malfatti, L., Bello, V., Mattei, G., Sada, C., Amenitsch, H., Gerdova, I., Hache, A., "PbS-Doped Mesostructured Silica Films with High Optical Nonlinearity." *Chemistry of Materials*, Vol. 17, 20, 2005, pp. 4965-4970.

[254] Li, Z., Hua, C., "Size tuning, functionalization and reactivation of Au in TiO2 nanoreactors." *Angewandte Chemie International Edition*, Vol., 44,28, 2005, pp. 4342-4345.

[255] Dekany, I., Turi, L., Galbacs, G., Fendler, J., "The effect of cadmium ion adsorption on the growth of CdS nanoparticles at colloidal silica particle interfaces in binary liquids." *Journal of Colloid and Interface Science*, Vol. 195, 2, 1997, pp. 307-315.

[256] Parvulescu, V., Constantin, C., Su, B., "Liquid phase oxidation of aromatic hydrocarbon using highly ordered Nb and NbCo-MCM41 nanoreactors." *Journal of Molecular Catalysis,* Vol. 202, 1-2, 2003, pp. 171-178.

[257] Ghadiri, M., Farzaneh, F., Ghandi, M., Alizadeh, M., "Immobilized copper (II) complexes on montmorillonite and MCM-41 as selective catalysts for epoxidation of alkenes." *Journal of Molecular Catalysis,*Vol. 233, 1-2, 2005, pp. 127-131.

[258] Masteri-Farahani, M., Farzaneh, F., Ghandi, M., "Synthesis and characterization of molybdenum complexes with bidendate Schiff base ligands within nanoreactors of MCM-41 as epoxidation catalysts." *Journal of Molecular Catalysis*, Vol. 248, 1-2, 2006, pp. 53-60.

[259] Shi, D., Li, R., Zhu, Y., Ke, Z., Yin, J., Jiang, W., Hu, G., "Nanoreactors for controlling the selectivity of free radical grafting of maleic anhydride onto polypropylene in the melt." *Polymer Engineering and Science,* Vol. 46,10, 2006, pp. 1443-1454.

[260] Shchukin, D., Svirdrov, D. "Photocatalystic processes in spatially confined micro-and nanoreactors." *Journal of Photochemistry and Photobiology*, Vol. 7, 1, 2006, pp. 23-39

[261] Zeng, H., "Ostwald Ripening: A synthetic approach for hollow nanomaterials." *Current Nanoscience*, 2007, Vol. 3,2, pp. 177-181.

[262] Shchukin, D., Sukhorukov, G., Moehwald, H., "Hollow micro- and nanoreactors for synthesis of new materials." *V. S. Physics, Chemistry and Application of Nanostructures: Reviews and Short Notes to Nanomeeting 2005*, May 24-27, 2005 (2005), pp. 493-496.

[263] Wingert, P., Mizukami, H., Ostafin, A., "Enhanced chemiluminescent resonance energy transfer in hollow calcium phosphate nanoreactors and the detection of hydrogen peroxide." *Nanotechnology*, Vol. 18,29, 2007, pp. 1-7.

[264] Li, J., Zeng, H., "Size tuning, functionalization, and reactivation of Au in TiO2 nanoreactors." *Angewandte Chemie, International Edition*, Vol. 44, 28, 2005, pp. 4342-4345.

[265] Zeng, H., "Synthetic architecture of interior space for inorganic nanostructures." *Journal of Materials Chemistry*, Vol. 16, 7, 2006, pp.649-662.

[266] Djojoputro, H., Zhou, X., Qiao, S., Wang, L., Yu, C., Lu, G., "Periodic mesoporous organosilica hollow sphere with tunable wall thickness." *Journal of the American Chemical Society*, Vol. 128, 19, 2006, pp. 6320-6321.

[267] Hah, H., Kim, J., Jeon, B., Koo, S., Lee, Y., "Simple preparation of monodisperse hollow silica particles without using templates." *Chemical Communications*, Vol. 14, 2003, 1712-1713.

[268] Chauhan, B., Latif, U., "Inorganic/Organic Hybrid Nanoreactors Based on Cyclic and Cubic Siloxane Scaffolds." *Macromolecules*, Vol. 38, 15, 2005, pp. 6231-6235.

[269] Wootton, R. Fortt, R., de Mello, A., "On chip generation and reaction of unstable intermediates-monolithic nanoreactors for diazonium chemistry: Azo dyes." *Lab on a Chip*, Vol. 2, 2002, 5-7.

[270] Bauer, S., Stock, N., "Functional porous materials. MOFs- metal organic frameworks." *Chemie in Unsere Zeit*, Vol. 42, 1, 2008, pp. 12-19.

[271] Kang, M., Lim, H., Lee, S., Lee, S., Kim, J., "Layered double hydroxide and its anion exchange capacity." *Advances in Technology of Materials and Materials Processing Journal*. Vol. 6, 2, 2004, pp. 218-223.

[272] Nikiforov, M., Chernysheva, M., Eliseev, A., Lukashin, A., Tretyakov, Y., Maksimov, Y., Suzdalev, I., Goernert, P., "Synthesis of iron oxide nanocomposites using layers double hydroxides." *Materials Science and Engineering B Solid Based Materials for Advanced Technology*, Vol. B109, 1-3, 2004, pp. 226-231.

[273] Papp, S., Dekany, I., "Growth of Pd nanoparticles on layer silicates hydrophobized with alkyl chains in ethanol-tetrahydrofuran mixtures." *Colloid and Polymer Science*, Vol. 280, 10, 2002, pp. 956-962.

[274] Sui, Y., Zhou, J., Skomski, R., Selimeyer, D., "Growth of Pd nanoparticles on layer silicates hydrophobized with alkyl chains in ethanol-tetrahydrofuran mixtures. Growth and magnetism of FePt:C composites in nanoscale channels." *Journal of Applied Physics*, Vol. 95, 11, 2004, pp. 6741-6743.

[275] Sui, Y., Zhou, J., Li, X., Skomski, R., Sellmyer, D., "Growth and magnetism of FePt:C composites in nanoscale channels." *Journal of Applied Physics*, Vol. 95, 11, 2004, pp. 6741-6743.

[276] He, J., "Preparation and thermal stability of gold nanoparticles in silk-templated porous filaments of titania and zirconia." *Chemistry of Materials*, Vol., 16,13, 2004, pp. 2656-2661.

[277] Lukashin, A., Vertegel, A., Eliseev, A., Nikiforov, M., Gornert, P., Tretyakov, Y., "Chemical design of magnetic nanocomposites based on layered double hydroxides." *Journal of Nanoparticle Research*, 2003, Vol. 5, 5-6, pp. 455-464.

[278] Lu, L., Capek, R., Kornowski, A., Gaponik, N., Eychmueller, A., "Selective fabrication of ordered bimetallic nanostructures with hierachical porosity." Vol., 44, 37, 2005, pp. 5997-6001.

[279] Dekany, I., Turi, L., Galbacs, G., Fendler, J., "Cadmium ion adsorption controls the growth of CdS nanopartiles on layered montmorillonite and calumite surfaces." *Journal of Colloid and Interface Science*, Vol. 213, 2, 1999, pp. 584-591.

[280] Torok, B., Bartok, M., Dekany, I., "The structure of chiral phenyethylammonium montmorillonites in ethanol-toluene mixtures." *Colloid and Polymer Science*, Vol. 277, 4, 1999, pp. 340-346.

[281] Epifani, M., Comini, E., Arbiol, J., Diaz, R., Sergent, N., Pagnier, T., Siciliano, P., Faglia, G., Morante, J., "Chemical Synthesis of In2O3 nanocrystals and their application in highly performing ozone sensing devices." *Sensors and Actuators*, Vol., 130,1,2008. 483-487.

[282] Mueller, A., Serain, C., "Soluble Molybdenum Blue-'des Pudels Kern,'" *Accounts of Chemical Research*, Vol. 33,1,2000, 2-10.

[283] Liu, T., "Surfactant induced trans-interface transportation and complex formation of giant polyoxomolybdate based clusters." *Journal of Cluster Science*, Vol. 14, 32003, pp. 215-226.

[284] Zhang, L., Shen, Y., Xie, A., Li, S., Wang, C. "One step synthesis of silver nanoparticles in self assembled multilayer films based on a Keggin structure compound." *Journal of Materials Chemistry*, Vol. 18, 11, 2008, pp. 1196-1203.

[285] Shchukin, D., Sukhorukov, G., Price, R., Lvov, Y., "Halloysite nanotubes as biomimetic nanoreactors." *Small*, Vol. 1, 5, 2005, pp. 510-513.

[286] Remskar, M., Mrzel, A., Virsek, M., Jesih, A., "Inorganic nanotubes as nanoreactors: the first MoS2 nanopods." *Advanced Materials*, Vol. 19,23, 2007, pp. 4276-4278.

[287] Valdes-Solis, T., Marban, G., Fuertes, A., "Preparation of nanosized perovshites and spinels through a silica xerogel template route." *American Chemical Society*, Vol. 17,8,2005, 1919, 1922.

[288] Wang, Z., Shang, H., Gil, U. "Nanoliter-scale reactor arrays for biochemical sensing." *Langmuir*, Vol. 22, 16, 2006, pp. 6723-6726.

[289] Vasylkiv, O., Borodianska, H., Sakka, Y., "Nanoreactor engineering and SPS densification of multimetal oxide ceramic nanopowders." *Journal of the European Ceramic Society*, Vol. 28, 5, 2008, pp. 919-927.

[290] Vasylkiv, O., Sakka, Y., Skorokhod, V., "Synthesis and properties of multimetal oxide nanopowders via nano-explosive technique." *Materials Science Forum*, Vol. 534-536, Pt. 1. Progress in Powder metallurgy. 2007, 125-128.

2

Miniemulsion Droplets as Nanoreactors

Katharina Landfester

The formulation and application of nanoparticle and nanocapsules composed of a polymeric, with or without an inorganic shell material, and a solid or liquid, inorganic or organic core, is of high interest for many applications in materials science. Many different approaches are used for the generation of polymeric nanoparticles and nanocapsules in order to obtain the demanded properties. For nanoparticles the emulsion polymerization is industrially usually used. For the formation of nanocapsules, the interfacial polymerization of a monomer or the interfacial nanodeposition of a preformed polymer is carried out [1]. Also, liposomes and block copolymers [2] can be used for encapsulation or the layer-by-layer deposition of polyelectrolytes [3] can be applied. In the liposome and copolymer approach, the shell material is hydrophilic and the units are not tightly bond which leads to unwanted leakage of the encapsulated material. For the other approaches, an adsorption on the surface is required, and the formation is kinetically driven.

In order to control the particle morphology, it takes advantage of a potential thermodynamic control for the design of nanoparticles, and the concept of "nanoreactors" where the essential ingredients for the formation of the nanoparticles are already in the beginning. It is underlined that nanoparticle and the nanocapsule formation in nanoreactors takes place in a highly parallel fashion (i.e., the synthesis is performed in 10^{18} – 10^{20} nanocompartments simultaneously which are separated from each other by a continuous phase). This continuous phase acts also as cooling reservoir for the reaction.

The idea of polymerization in a nanoreactor is technically realized in high perfection in the suspension polymerization, where droplets in the micrometer range are created which can be polymerized without changing the particle identity [4]. Here micrometer large particles are obtained. The suspension principle was transferred to obtain smaller droplet sizes by Ugelstad [5] who scaled down the droplet size by applying high shear to a heterophase system to several hundred nanometers by shearing the system.

It is the topic of this chapter to describe a recent development where the availability of high-shear devices such as ultrasound and high-pressure homogenizers in combination which some applied physical chemistry has decreased the droplet or nanoreactor diameter down to 30 to 100 nm, still preserving the integrity of each single nanodroplet. It also appears that this developed concept is not restricted to a single procedure (such as radical polymerization in water), but turned out to be widely applicable to generate nanocapsules via a liquid/liquid technology both in reverse (aqueous solvent) and inverse (organic or hydrocarbon solvent) situations.

A system where small and narrowly distributed droplets with high stability and a diameter between 30 nm and 500 nm in a continuous phase are created by using high-shear (ultrasonifier or high-pressure homogenizers) [6–8] is classically called a "miniemulsion." One of the tricks to obtain a high stability of the droplets is the suppression of the Ostwald ripening which can be achieved by the addition of an agent which dissolves in the dispersed phase, but is insoluble in the continuous phase. This agent cannot diffuse from one droplet to the other and is trapped in each droplet; this provides an osmotic pressure inside the droplets which counteracts the Laplace pressure. The effectiveness of the hydrophobe increases with decreasing water solubility in the continuous phase. This mechanism was already used for the stabilization of fluoroalkane droplets by addition of perfluorodimorphineopropane, which results in an effective and stable blood substitute [9].

Because of the high stability of the droplets, each miniemulsion droplet can indeed be treated as a small nanoreactor. This enables a whole set of new polymerization reactions that lead to nanoparticles that were not accessible before; radical polymerization, anionic, cationic, enzymatic, catalytic polymerization, polycondensation, polyaddition, and oxidative polymerization can be carried out in the small nanoreactors. Some examples will be given in the following sections to show its wide applicability, for example in materials science, biomedicine, pharmaceutics, and cosmetics.

For a typical oil-in-water miniemulsion, an oil, a hydrophobic agent (or several), an emulsifier, and water are homogenized by high shear (see Figure 2.1) to obtain homogeneous and monodisperse droplets in the size range of 30 to 500 nm. The polymerization in the droplets can be achieved by a subsequent reaction.

2.1 Different Kinds of Polymerization in the Nanoreactors

The miniemulsion process allows in principle the use of all kinds of different monomers for the formation of particles, which are not miscible with the continuous phase. In case of prevailing droplet nucleation or the start of the polymer reaction in the droplet phase, each miniemulsion droplet can indeed be treated as a small nanoreactor. This enables a whole variety of polymerization reactions that lead to nanoparticles (much broader than in emulsion polymerization) as well as to the synthesis of nanoparticle hybrids, which were not accessible before.

2.1.1 Radical Polymerization

As a model monomer for radical homopolymerization of hydrophobic monomers, styrene is described in many papers. The polymerization of acrylates and methacrylates is also well known. It could be also shown that

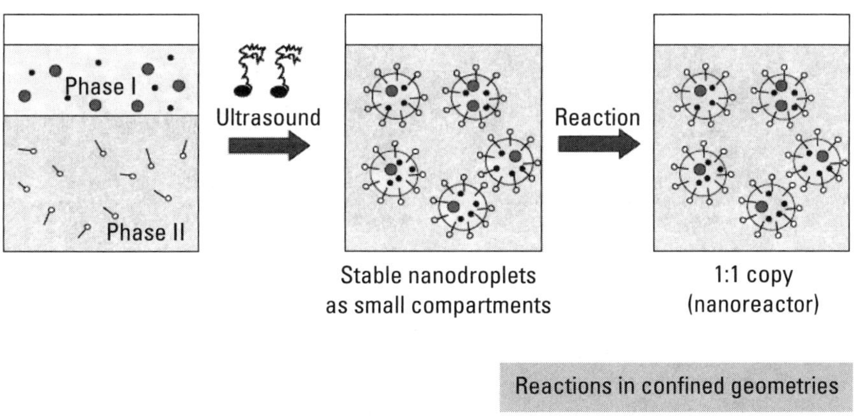

Figure 2.1 Principle of the miniemulsion process.

the miniemulsion process easily allows the polymerization of the ultrahydrophobic monomer lauryl methacrylate [see Figure 2.2(a)] or isobornyl acrylate [10] without any carrier materials as necessary in emulsion polymerization. Not only simple hydrophobic, but also fluorinated water-insoluble monomers were applied for the synthesis of latexes in the size range of 100 to 250 nm by employing rather low doses of usual nonfluorinated surfactants [11].

Of high interest are semicrystalline polyacrylonitrile particles. A peculiarity of polyacrylonitrile is that it is insoluble in its monomer which makes it very difficult to homopolymerize acrylonitrile in an emulsion polymerization process since nucleated polymer particles cannot grow by monomer swelling. Polymerization in miniemulsion, however, is a very suitable technique to avoid this problem since each droplet acts as a nanoreactor. As a

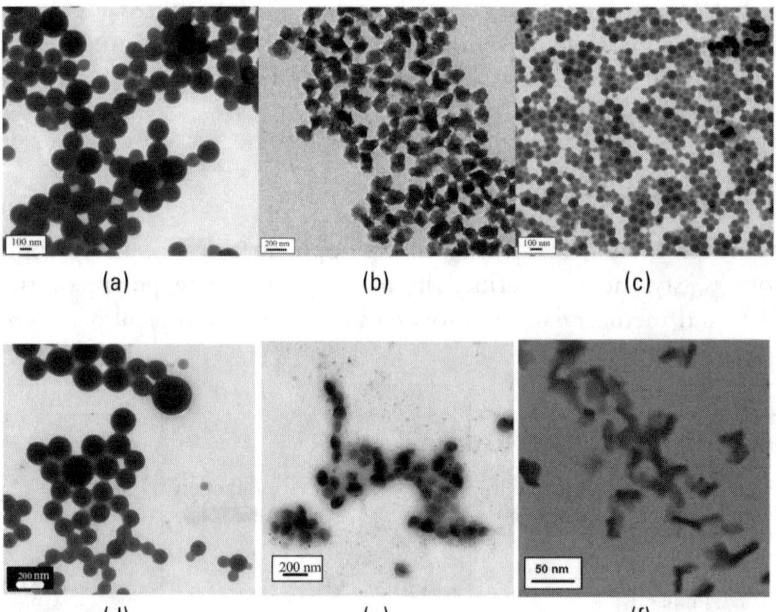

Figure 2.2 TEM photographs of (a) laurylmethyacrylate latex particle; (b) polyacrylonitrile latex particles obtained by the direct (radical) miniemulsion polymerization; (c) hydrophilic polyacrylamide particles obtained in inverse radical miniemulsion polymerization; (d) polyurethane particles obtained by polyaddition in miniemulsion; (e) polyester particles obtained by enzymatic polymerization; and (f) poly-ε-caprolactam (polyamide-6) by ring-opening anionic polymerization in miniemulsion.

result, pure polyacrylonitrile nanoparticles (PAN) were obtained [see Figure 2.2(b)] [12]. Due to the insolubility of the polymer in the monomer, the formed polymer precipitates and crystallizes during the polymerization within the droplets; about 10-nm-large polymer nanocrystals are formed. The polymerization of more hydrophilic monomers is also possible, as shown for methyl methacrylate (MMA) and vinyl acetate [13–15]. In the case of such monomers with a pronounced water solubility, the nucleation in water should be efficiently suppressed in order to avoid secondary nucleation in the water phase. This can be achieved (e.g., by using an oil-soluble initiator). The use of allylmethyl carbonate and di(ethylene gycol)bisallylcarbonate allows one to obtain polycarbonate particles in a radical miniemulsion polymerization using benzoyl peroxide as a effective initiator [16].

It was shown that the principle of aqueous miniemulsions could be successfully transferred to nonaqueous media [17]. Here, polar solvents, such as formamide or glycol replace water as the continuous phase, and hydrophobic monomers are miniemulsified with a hydrophobic agent, which stabilizes the droplets against molecular diffusion processes. In the case of inverse systems, hydrophilic monomers such as hydroxyethyl acrylate, acrylamide [particles see Figure 2.2(c)], and acrylic acid were miniemulsified in nonpolar media (e.g., cyclohexane or hexadecane) [17, 18]. Recently redox initiation system consisting of ceric ions and carbohydrate-based surfactant Span 60 as a reducing agent was successfully used for the polymerization of AAm [19].

The miniemulsion is also well suited for the preparation of copolymers. Here, a mixture of different hydrophobic monomers can be used (e.g., a mixture of styrene and methyl methacrylate [20], styrene and butyl acrylate [21], butyl acrylate and vinyl acetate [22, 23], or styrene and butadiene [24, 25]). MMA was shown to copolymerize with butyl acrylate [26], or with more hydrophobic monomers like p-methylstyrene, vinyl hexanoate, or vinyl 2-ethylhexanoate, where one of the monomers acts as the hydrophobe [27]. The resulting copolymer composition tended to follow the predictions of the reactivity ratios (i.e., the reaction progresses such as a bulk reaction). Fluorinated monomers can also be copolymerized with monomers as MMA and styrene. Alternatively, fluoro-containing latexes could also be obtained by miniemulsion polymerization of styrene and n-butyl methacrylate using a fluorinated surfmer which in incorporated in the polymer chain and is located at the outside of the particles resulting after film formation in low free-energy surfaces [28]. In order to obtain hydrophobic and oleophobic latexes, polysiloxane-acrylate latexes with small and narrowly distributed particle sizes have been synthesized [29]. The polysiloxane acrylate latexes were obtained in a radical polymerization process; here a copolymerization with other vinylic monomers is easily possible leading to highly crosslinked copolymer particles. Due to the confinement of the reaction within the minidroplets, the monomers are forced to copolymerize on the length scale of the droplets and the phase separation is limited on the size of the particles. Not only the combination of two hydrophobic monomers can be used, but also a hydrophobic and a hydrophilic monomer can be copolymerized. Small amounts of hydrophilic monomers as acrylic acid (AA) or methacrylic acid (MAA)) [30] or 2-hydroxyalkyl methacrylates [31] could be easily used in a styrene miniemulsion polymerization. The polymerization process of two monomers with different polarities in similar ratios is a difficult task due to the solubility problems. Using the miniemulsion process, it was possible to start from very different spatial monomer distributions, resulting in very different amphiphilic copolymers in dispersion [32]. The monomer, which is insoluble in the continuous phase, is miniemulsified in order to form stable and small droplets with a low amount of surfactant. The monomer with the opposite hydrophilicity dissolves in the continuous phase (and not in the droplets).

In a similar way, the hydrophilic cationic monomer [2-(methacryloyoxy)ethyl]trimethyl ammonium chloride (MAETAC) was copolymerized with the hydrophobic butyl acrylate using cumene hydroperoxide/tetraethylene pentamine as a redox initiator system in order to obtain amphiphilic polymer nanoparticles [33]. The copolymerization with such functional hydrophilic monomers allows a prefunctionalization of the nanoparticles, creating carboxylic or amino containing surfaces where well-defined surface charges can be created. A final functionalization can be obtained by binding amino acids, peptides, or antibodies onto the prefunctionalized nanocapsules in order to achieve a high selectivity (e.g., for the uptake of the particle in different cell lines) [34]. Polystyrene-block-poly(methyl methacrylate) block copolymers were produced in situ by controlled radical polymerization (CRP) through the addition of the second monomer to a seed prepared by miniemulsion polymerization with a certain amount of a CRP agent [35]. With an increase in the amounts of the block copolymers, it was shown that the particle morphology changed from a hemisphere morphology (for a latex without block copolymers (i.e., without the use of a CRP agent during the polymerization) to clear core-shell morphologies as a result of decreasing polymer-polymer interfacial tension.

Waterborne poly(urethane-block-styrene) latexes were prepared by miniemulsion polymerization in only one batch [36]. A direct miniemulsion of the monomers mixture containing styrene, isophorone diisocyanate, 2,4-diethyl-1,5-pentanediol and a diol-functionalized azoinitiator was prepared in water. The polyaddition reaction of the polyurethane was then performed in a first step at room temperature, with the help of the catalyst mixture dibutyltindilaurate/dimethyldodecylamine, yielding a polyurethane macroazoinitiator. In a second step, the miniemulsion was heated to 72°C to start the radical polymerization of styrene from the macroazoinitiator chains. It was shown that 45% of a (linear) copolymer consisting of a polyurethane and polystyrene was obtained, which is a good compatibilizer for polyurethane/polystyrene polymer blends. Transmission electron microscopy (TEM) from the polymer particles in miniemulsion revealed a homogeneous structure inside the particles.

Already preformed polymers were used for the formation of particles for optical applications [37–40]. For the elucidation of the morphology, polymer blends in nanoparticles have been studied by TEM and photoluminescence (PL) spectroscopy [41]. The TEM studies show that blend particles formed from two immiscible polymers by the miniemulsion process exhibit biphasic morphologies. The fact that no core-shell type but Janus-like structures were found indicates that the surface-free energies between both polymers and the solution-water interface (including the surfactant molecules) are similar; therefore, the blend morphology and composition of the individual phases are mainly determined by the interaction between the two polymer components. Both the TEM studies and the PL experiments provide strong evidence that phase separation in these particles strictly follows the Flory-Huggins theory. This highlights the applicability of the nanoparticle approach to fabricate blend systems with well-controllable properties and to study structure-property relationships under well-defined conditions.

2.1.2 Controlled Free-Radical Miniemulsion Polymerization

Living free-radical polymerization represents a promising technique to produce polymers with highly controlled structures. Different possible systems known from bulk polymerizations have been used in miniemulsions. The living free-radical polymerization of (e.g., styrene via the miniemulsion approach) allows one to eliminate the drawback of the bulk system where an increase in polydispersity was found at high conversions due to the very high viscosity of the reaction medium [42]. Different approaches for controlled radical polymerization have been adapted to the miniemulsion polymerization process. In a stable free-radical polymerization (SFRP), the initiated polymer chains are reversibly capped by a stable radical, in the case of the nitroxide-mediated SFRP by a stable nitroxide like the 2,2,6,6-

tetramethylpyridin-1-oxyl radical (TEMPO). Stable PS dispersions via miniemulsion polymerization are prepared by MacLeod et al. with an optimized ratio and amount of surfactant, hydrophobe, nitroxide, and KPS as initiator at 135°C [43]. TEMPO can also be used in combination with benzoyl peroxide (BPO) [44], whereas the use to KPS initiation resulted in more chains of lower molecular weight than BPO [45]. Recently, nitroxide-mediated polymerization of styrene has demonstrated that it can be performed in a continuous tubular [46]. It was shown that the polymerization kinetics in the tubular reactor are similar to those in a batch reactor. The number average molecular weight increases linearly with conversion, and chain extension experiments were successful, indicating that the living nature of the polymerization is maintained in the tubular reactor. With the use of an acrylic β-phosphonylated nitroxide, the N-tert-butyl-N-(1-diethylphosphono-2,2-dimethylpropyl) nitroxide, and a water-soluble alkoxyamine as the initiator [47], faster kinetics than those with TEMPO were observed which allows a reduction of the reaction temperature. More importantly, this nitroxide was shown to be particularly well suited for the controlled polymerization in miniemulsion of acrylic esters such as butyl acrylate [48] and allowed also the preparation of defined poly(n-butyl acrylate-co-styrene) gradient copolymers [49] or diblock copolymers [50].

The TEMPO-mediated stable free-radical polymerization of styrene in miniemulsion could also be performed at 100°C (and therefore in nonpressurized reactors) with reasonable reaction rates by the addition of ascorbic acid or a free-radical initiator [51]. It was shown that the living character of the chains was preserved; the degree of polymer "livingness" was comparable to polymerizations conducted at 135°C. Polydispersities were broader than those observed in well-controlled systems, ranging from similar to 1.4–1.6, and consistent with expectations for systems having a low activation rate.

Controlled radical polymerizations in miniemulsions have also been conducted by de Brouwer et al. using reversible addition-fragmentation chain transfer (RAFT) and nonionic surfactants [52]. The polydispersity index was usually below 1.2. The "living" character is further exemplified by its transformation into block copolymers. In order to restrict all polymerization reactions better to within the miniemulsion droplets, a preceding bulk step in which dithiobenzoate-end-capped oligomers were synthesized, followed by conducting the second part after emulsification of the resultant oligomers [53]. Stability-enhanced functioanlized latexes could be obtained by using a RAFT agent bearing a carboxylic acid group in the miniemulsion polymerization of methyl methacrylate [54]. A xanthate-mediated RAFT polymerization in miniemulsion has been shown to facilitate the controlled radical polymerization of vinyl acetate in miniemulsion [55]. Two trithiocarbonate RAFT agents were used in miniemulsion polymerization of styrene and butyl acrylate and the formation of seeded emulsion block copolymers. The type of surfactant in relation to the monomer used was found to have a significant affect on latex formation. Conditions are shown by which AB and ABA-type block copolymers can be successfully prepared via a seeded RAFT-mediated emulsion polymerization [56]. The RAFT process could also be performed as continuous miniemulsion block copolymerization of styrene and n-butyl acrylate in a train of continuous stirred tank reactors (CSTRs). It was shown that a train of CSTRs can effectively be used to produce unique multiblock copolymers [57]. The polydispersity of the polymer decreased as the number of CSTRs in the train increases [58]. A continuous RAFT miniemulsion polymerization was even performed in a tubular reactor which shows the feasibility of this process on an industrial basis [59].

Reverse atom transfer radical polymerization (ATRP) of butyl methacrylate was successfully conducted in miniemulsions by Matyaszewski et al. using the water-soluble initiator V50 and the hydrophobic ligand 4,4'-di(5-nonyl)-4,4'-bipyridine (dNbpy) to complex the copper ions. Although the forming radical mediator Cu(II) complex had a large water-partitioning coefficient, the rapid transfer of Cu(II) between the organic and aqueous phases assured an adequate concentration of the deactivator in the organic phase [60, 61]. The development of highly active transition-metal complexes with hydrophobic ligands allows the preparation of polymers with well-defined compositional topological control [62]. A miniemulsion ATRP via activators generated by electron transfer (AGET) starting from an oxidatively stable catalyst system was shown to be successful for the preparation of linear and star-shaped block copolymers [63]. An excess of reducing agents as ascorbic acid and tin(II) 2-ethylhexanoate consumes the oxygen present in the system and, therefore, provides a deoxygenated environment for ATRP. This allowed ATRP of butyl acrylate in miniemulsion and in the presence of air to be successfully carried out. During polymerization the radical concentration remained constant and the polymerization reached over 60% monomer conversion after 6 hours [64]. Well-defined linear and star-shaped block copolymers were synthesized using halogenated ATRP macroinitiator via a simultaneous reverse and normal initiation ATRP process in both bulk and stable aqueous miniemulsion [65]. For targeted drug delivery scaffolds in biomedical application, stable biodegradable nanogels crosslinked with disulfide linkages were prepared by inverse miniemulsion ATRP [66]. The uniformly crosslinked network allows improved control over the release of encapsulated agents. The nanogels can biodegrade into water-soluble polymers in the presence of a biocompatible glutathione tripeptide, which is commonly found in cells, and therefore release encapsulated molecules including rhodamine 6G, a fluorescent dye, and the anticancer drug doxorubicin.

2.1.3 Anionic Polymerization

For the anionic polymerization of phenyl glycidyl ether (PGE) in miniemulsion, Maitre et al. used didodecyldimethylammonium hydroxide as an inisurf, which acts as a surfactant and an anionic initiator by means of its hydroxy counterion at the same time [67]. The ring-opening polymerization of 2,4,6,8-tetramethylcyclotetrasiloxane in miniemulsion, using mono- and diphosphonic-acid surfactants generates in less than 15-min linear poly(methylhydrogenosiloxane) (PMHS) of large molar masses (typical number-average of 25 kg/mol), of low polydispersities (around 2), and in good yields (up to 94%) [68]. Molar masses can be tuned at will.

The analysis of homopolymerizations of octamethylcyclotetrasiloxane and tetramethyltetravinylcyclotetrasiloxane in miniemulsion focusing specifically on the rates of polymerization, backbiting, and polycondensation processes [69] showed that the advantage of working in dispersed media is that different scales of reactivities are present between one-chain reactions, taking place at the interface (propagation, backbiting), and two-chain reactions, located in the bulk of the particles (polycondensation and intermolecular redistribution).

Vinyl functionalized multiblock silicones could also be obtained by the process of anionic ring opening polymerization in miniemulsion. Copolymerization of octamethylcyclotetrasiloxane with tetramethyltetravinylcyclotetrasiloxane gave insight into the extent of mixed cycles' formation and particularly how to avoid these. On this basis, recipes were carefully selected so that homopolymerization and scarce copolycondensation and redistribution progress at the expense of backbiting reactions and thus multiblock copolymers are formed, the microstructure of which was confirmed with Si-29 NMR [69]. Rehor et al. applied the anionic ring-opening polymerization of episulfides in emulsion. The polymerization proceeded with a living mechanism, but was characterized by a limiting conversion, presumably arising from the increase in viscosity in the polymer [70].

The miniemulsion process can be also applied for the preparation of poly(n-butylcyanoacrylate) (PBCA) nanoparticles [71]. In the first step a miniemulsion is prepared from n-butylcyanoacrylate in hydrochloric acid solution using sodium dodecylsulfate as surfactant. In the second step, a base solution is added to initiate polymerization and the polymeric particles are formed. Using amines or amino acids as initiators allowed the convenient functionalization of the polymer particles' surface. The influence of surfactant concentration and sonication time on particle size and size distribution has been studied as well as the influence of pH, concentration, and amount of initiator on the particle size and the distribution of the molar mass of the polymer. Amino acid and MethoxyPEG functionalization could be introduced by using aqueous solutions as initiator for the anionic polymerization in heterophase [72]. All prepared particles have sizes smaller than 250 nm and negative zeta potentials. The molar mass distribution of the polymer is dependent on the acid used as continuous phase and the applied initiator solution. Cells of three lines (HeLa, Jurkat, mesenchymal stem cells) were incubated with the particles. The molar mass of the polymer determines the onset and extent of apoptosis, the total uptake is determined by the size and functionalization of the particles. Recently, it was shown that nanoparticles of polyamide 6 could be synthesized using the inverse miniemulsion technique [73]. Even though stable molten ε-caprolactam droplets could be created at low temperatures, these droplets showed a limited stability at the reaction temperature of 150°C. Therefore, the monomer ε-caprolactam, the initiator NaH, and the classical activator N-acetylcaprolactam was dissolved in DMSO in order to build the polar phase, which was stably after dispersing in the inexpensive isoparaffinic continuous phase also at 150°C. During the polymerization, a demixing of the DMSO and the polymer occurred and the polymer precipitated inside the DMSO droplets leading to anisotropic nonspherical nanoparticles as revealed by TEM [see Figure 2.2(f)]. The polyamide 6 exhibited a relatively high molecular weight of 35,000 to 40,000 g·mol–1 as determined by viscosimetry measurements. The absence of structural defects, that could lead to less good mechanical properties of the polymer, could also be shown.

2.1.4 Cationic Polymerization

The cationic polymerization of p-methoxystyrene in miniemulsion could successfully be performed by using an acidic initiator and ytterbium triflate as a cocatalyst. Since ytterbium triflate dissociates in water leading to a high ionic strength, the use of an electrosteric surfactant (i.e., sodium dodecylpolyoxyethylene sulfate) is required. The catalyst increases significantly the polymerization rate, but only moderately affects the molar masses [74]. If the polymerization is exclusively carried out inside the droplets preventing fast transfer reactions with water, the generation of poly(p-methoxystyrene) with a molar mass of several thousand g·mol^{-1} is described [75]. The cationic polymerization of styrene was also performed by using B(C6F5)(3) as a water-tolerant Lewis acid [76].

2.1.5 Enzymatic Polymerization

The enzymatic reaction in bulk or solution has the drawback that a high amount, about 50 wt% of enzyme, is needed compared to the monomer, that the conversion is only about 80% even after a long reaction of up to 5 days and that the formed polymers only show low molecular weight. The problem is that the enzyme can not be efficiently active in a hydrophobic environment. Therefore, the direct enzymatic polymerization of miniemulsions consisting of lactone nanodroplets in water represents a new and convenient pathway for the synthesis of biodegradable polymer nanoparticles [see Figure 2.2(e)], where the chemical composition and molecular weight can be varied in a certain range. Due to the amphipilicity of the enzyme (lipase PS), the reaction takes place at the particle/water interphase. Within a short reaction time of about 2 hours, 100% conversion is achieved and particles with a high molecular weight of about 200,000 g/mol are obtained. Oligoesters completely end-capped by an alkene or diene group can also be made by this technique. These building blocks expectedly extend polyester application as they allow the carrying of improved biodegradability, both to siloxane and resin chemistry [77].

2.1.6 Oxidative Polymerization

Polyaniline (PANI) latex particles have been synthesized using both inverse and direct miniemulsion polymerization techniques [78]. In the case of inverse miniemulsion, aniliniumhydrochlorid was dispersed and could be oxidized by simple H_2O_2, thus resulting under optimized conditions in highly crystalline PANI particles dispersed in hydrocarbons with high-volume fraction. The resulting structure is identified as emeraldine which, however, crystallizes in a new crystal morphology. Aniline itself can be used for the formulation of direct miniemulsion using typical surfactants, such as the ionic surfactants SDS, CTMA-Cl, the nonionic surfactant Lutensol AT50, and the block copolymeric surfactant SE3030. Oxidation of aniline miniemulsions in water with diverse oxidants from the continuous phase also leads to highly crystalline PANI, which, however, has a low colloidal stability. Additional employment of a costabilizer like PVA or PVP in this case leads to stable aqueous PANI dispersions for water-based systems otherwise unreached by local structural order.

2.1.7 Catalytic Polymerization

Ethylene can successfully be polymerized in catalytic emulsion polymerization [79]. To obtain polymer latexes, water-soluble complexes can be employed as catalysts. If hydrophobic catalysts or water-sensitive catalyst should be used, aqueous miniemulsion consisting of the lipophilic catalyst or catalyst precursors dissolved in a small amount of hydrocarbon can be prepared. As catalysts, Ni(II) complexes were used in order to obtain high molecular weight polyethylene [80]. It is even possible to use miniemulsions of in situ catalysts which are based entirely on compounds commercially available from standard chemicals suppliers, namely a chlorinated derivative of 1,4-benzoquinone, triphenylphosphine, and bis(1,5- cyclooctadiene)nickel. The catalyst is stable in water for hours with productivities of up to $2 \cdot 10^3$ mol (ethylene)·mol(Ni)$^{-1}$·h^{-1}. These productivities and molecular weights are in the same range as conventional polymerization in organic solvents [81]. Using a cobalt catalyst system in a miniemulsion, crystalline syndiotactic 1,2-polybutadiene could also be obtained [82]. The employment of a catalyst miniemulsion with palladium(II) catalysts allows also the copolymerization of an olefin with CO in order to obtain aqueous polyketone latex particles [83]. Polystyrene nanoparticles could also be used as organic supports for olefine polymerization in miniemulsion by using metallocenes as catalysts. The support particles are functionalized with nucleophilic surfaces such as polyethylenoxide, polypropyleneoxide, or pyridine units allowing an immobilization of the metallocene catalysts via a noncovalent immobilization process. Remarkably, high activities and excellent product morphologies were obtained [84]. Chemtob and Gilbert implemented the catalytic insertion polymerization of norbornene under miniemulsion conditions. Using sodium dodecyl sulfate as surfactant and hexadecane as hydrophobe, the insertion polymerization was initiated by multicomponent catalysts generated in situ by reacting allylpalladium chloride dimer, with a desired phosphine ligand and then activating the resulting procatalyst with a salt of noncoordinating anion, typically the lithium salt of tetrakis(pentafluorophenyl)borate. Stable polynorbornene latexes with a relatively broad particle size distribution were produced. The number-average particle diameter is often below 100 nm [85]. Both polynorbornene and polybutadiene-based particles were prepared by the respective ring-opening metathesis polymerization of norbornene and cyclooctadiene initiated by

(PCY3)2Cl2Ru=CHPh using also the miniemulsion process resulting in particles of about 300 nm [86]. Air-stable atalytic surfactants could serve as inisurf molecules for the ring-closing metathesis polymerization in water [87].

2.1.8 Polyaddition Reaction

As already indicated in the introduction, the existence of stable, isolated nanodroplets, in which chemical reactions may, but do not have to, depend on droplet exchange (the so-called nanoreactors), enables the application of the miniemulsion process in a much broader range. Contrary to the process of creating a secondary dispersion as it was used for the preparation of (e.g., polyurethanes and epoxide resins), it was shown that the miniemulsion polymerization process allows one to mix monomeric components together, and polyaddition and polycondensation reactions can be performed after miniemulsification in the miniemulsificated state [88]. The principle of miniemulsion polymerization to polyadditions of epoxy-resins was successfully transferred to mixtures of different epoxides with varying diamines, dithiols, or diols which were heated to 60°C to form the respective polymers [88]. The requirement for the formulation of miniemulsions is that both components of the polyaddition reaction show a relatively low water solubility, at least one of them even below 10^{-5} $g{\cdot}L^{-1}$. The final polymers reveal molecular weights of about 20,000 $g{\cdot}mol^{-1}$ with a disparity of close to 2. This means that unexpectedly ideal reaction conditions are preserved during the reaction in miniemulsion, and that the proximity of the interface to water does not really disturb the reaction. As it was shown that also polyurethane latexes can be made by direct miniemulsification of a monomer mixture of diisocyanate and diol in an aqueous surfactant solution followed by heating; see Figure 2.2(d) [89]. This is somewhat special since one might expect a suppression of polymerization by side reactions between the very reactive diisocyanates and the continuous phase water. Therefore, it is important that the reaction between diisocyanate and diol has to be slower than the time needed for the miniemulsification step and the side reaction of the diisocyanate with water in the dispersed state has to be slower than the reaction with the diol. Numerous parameters permitted to decrease the proportion of this side reaction. Among them, the use of organo-tin compounds allowed almost triple the polyurethane molecular weight [90]. Also, the knowledge of the localization of the reaction has been used for this purpose. For example, by increasing the particle size, the isocyanate-water reaction, which takes place at the interface of the particle, was reduced compared to the isocyanate-hydroxyl reaction, which is located in the core. Also, by increasing the amount of dodecanediol located at the interface, the reaction with water decreased and the molecular weight increased. Without any extensive optimization, molecular weight in the range of 70,000 $g{\cdot}mol^{-1}$ could be obtained which is very promising regarding the potential of the system. Polysiloxane urethanes can be obtained in a polyaddition process using a diisocyanate and silanediols. Replacing parts of the low glass transition temperature siloxanediol segments by alkyldiols allows one to introduce crystalline parts with higher melting points in the material [29]. These alkyldiols however have to be sufficiently hydrophobic that they do not interfere with the miniemulsion process.

2.1.9 Polycondensation Reaction

Even though it seems to be a contradiction, hydrophobic polyesters have been synthesized in miniemulsion in the presence of large amounts of water [91]. The yield of the esterification and the molecular weight of the polyesters have been determined for different reaction conditions. It was found that the dispersion state has no influence on the equilibrium: the yield is the same in 100-nm particles than in very large droplets. However, an important parameter is the water concentration inside the particles, as shown by an increase of the yield with more hydrophobic monomers. Another important parameter is the structure of the alcohol monomer. Alcohol compounds bearing electron-donating groups displace the equilibrium toward ester formation.

2.1.10 Polymerase Chain Reaction

Polymerase chain reaction (PCR) was successfully carried out inside stable and narrowly distributed water-in-oil nanodroplets with a size of 100 to 300 nm in diameter. The droplets were obtained by the miniemulsion process [92]. Each aqueous droplet serves as a single nanoreactor for the PCR. It was found that the size of the droplets highly depends on the sonication parameters (i.e., time and amplitude) and that these parameters have a great influence on the final concentration of the PCR product. The parameters were chosen the way that conditions for single molecule chemistry were obtained, since the three-dimensional space is compartmentalized in small nanoreactors in each of which the same reaction takes place in a highly parallel fashion on every single DNA molecule.

2.2 Formation of Nanocapsules

2.2.1 Generation of Encapsulated Inorganics

It is of course of high interest to combine an inorganic and the polymeric part in order to obtain nanocapsules. If the inorganic particles are dispersed in a monomer and if this dispersion undergoes a miniemulsification process, polymeric particles with fully encapsulated inorganic material are obtained. The polymeric shell protects the inorganic particle efficiently. It is possible to incorporate just one (large) inorganic particle per polymer particle (e.g., $CaCO_3$) [93], or one can incorporate many small particles (e.g., hydrophobized magnetite particles), see Figure 2.3(b) [94].

Since carbon black is a rather hydrophobic pigment (depending on the preparation conditions), the encapsulation of carbon black in the latexes by direct dispersion of the pigment powder in the monomer phase prior to emulsification is again a suitable way [93]. Here, full encapsulation of nonagglomerated carbon particles can be provided by the appropriate choice of the hydrophobe. In this case the hydrophobe not only acts as the stabilizing agent against Ostwald ripening for the miniemulsion process, but also mediates to the monomer phase by partial adsorption. However, this direct dispersion just allows the incorporation of 8 wt% carbon black since the carbon is still highly agglomerated in the monomer. At higher amounts, the carbon cluster broke the miniemulsion, and less defined systems with encapsulation rates lower than 100%, which also contained pure polymer latexes, were obtained. To increase the amount of encapsulated carbon to up to 80 wt%, another approach was developed [95] where both monomer and carbon black were independently dispersed in water using SDS as a surfactant and mixed afterwards in any ratio between the monomer and

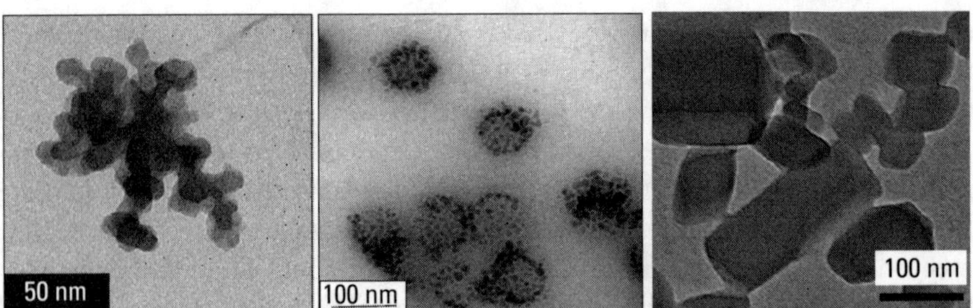

Figure 2.3 Incorporation of inorganic materials: (a) carbon black, (b) magnetite particles, and (c) azo pigment particles.

carbon. Then, this mixture was cosonicated, and the controlled fission/fusion process characteristic for miniemulsification destroyed all aggregates and liquid droplets, and only hybrid particles being composed of carbon black and monomer remain due to their higher stability [Figure 2.3(a)] [95]. This controlled droplet fission and heteroaggregation process can be realized by high-energy ultrasound or high-pressure homogenization.

Composites of a styrene-isoprene copolymer and single wall carbon nanotubes (SWNTs) could be also prepared by using the process of miniemulsion polymerization [96]. Here two strategies were followed: In the first procedure, the polymerization took place in the presence of SWNTs leading to an aggregation of SWNTs. Mixing dispersed SWNTs with latex after reaction was able to preserve the SWNT dispersion and gave a polymer composite with an electrical percolation threshold of 0.2%. In order to provide different colors for ink jet printing, different organic yellow, magenta, and blue pigment nanoparticles were efficiently encapsulated by cosonicating a pigment dispersion and a monomer miniemulsion in a weight ratio pigment to monomer of 80:20, followed by polymerization [97]. Efficient encapsulation was proven by means of ultracentrifugation, electron microscopic methods [particles see Figure 2.3(c)] and streaming potential titration. Comparing the reaction kinetics of a typical styrene miniemulsion polymerization with corresponding "ad-miniemulsion polymerizations" proceeding on the surface of pigment particles, an influence of the pigments' molecular structure was observed. In order to obtain different charged particle surfaces, the encapsulation process was performed using anionic, cationic, or nonionic surfactants. Both their amount and type have a large influence on the encapsulation process. However, not only the surfactant but also the monomer can be varied.

Changing now the hydrophilicty of the encapsulating material from hydrophobic to hydrophilic, the material has to be hydrophobized prior to the encapsulation process. The encapsulation of hydrophilic magnetite particles into polystyrene particles was efficiently achieved by a miniemulsion process using oleoyl sarcosine acid [98] or the more efficient oleic acid as first surfactant system to handle the interface magnetite/styrene, and SDS to stabilize the interface styrene/water, thus creating a polymer-coated ferrofluid. The encapsulation of high amounts of magnetite into polystyrene particles can efficiently be achieved by a new three-step preparation route including two miniemulsion processes; see Figure 2.3(b) [99, 100]. In the first step, a dispersion of oleic acid coated magnetite particles in octane is obtained. In the second step, magnetite aggregates in water are produced in a miniemulsion process by using sodium dodecyl sulfate (SDS) as surfactant.

In the third step, the dispersion with the magnetite aggregates, which are covered by an oleic acid/SDS bilayer, were mixed with a monomer miniemulsion and a second miniemulsion process, an ad-miniemulsification process, is used to obtain full encapsulation. After polymerization, polymer encapsulated magnetite aggregates were obtained. Such particles can be used for biomedical applications (e.g., detection by magnetic resonance tomography and destruction of tumor cells by hyperthermia).

In order to obtain "dual reporter particles" magnetite and a fluorescent dye were encapsulated in polystyrene nanoparticles [101]. The polymerization of the monomer styrene yielded nanoparticles in the range of 45 to 70 nm. By copolymerization of styrene with the hydrophilic acrylic acid, the amount of carboxyl groups on the surface was varied. For biomedical evaluation, the nanoparticles were incubated with different cell types. The introduction of carboxyl groups on the particle's surface enabled the uptake of nanoparticles as demonstrated by the detection of the fluorescent signal by fluorescent-activated cell sorter (FACS) and laser-scanning microscopy. The quantity of iron in the cells that is required for most biomedical applications (like detection by magnetic resonance imaging) has to be significantly higher as can be achieved by the uptake of magnetite encapsulated nanoparticles functionalized only with carboxyl groups. A further increase of uptake can be accomplished by transfection agents like poly-L-lysine or other positively charged polymers. This functionality was also engrafted into the surface of the nanoparticles by covalently coupling of lysine to the carboxyl groups. The amount of iron that can be transfected was even higher than with the nanoparticles with a transfection agent added and by this only physically adsorbed. Furthermore, the subcellular localization of these nanoparticles was demonstrated to be clustered in endosomal compartments.

CdSe/ZnS core-shelf quantum dots (QDs) could be also used for encapsulation purposes [102]. It was shown that the polymerization kinetics were not altered by the presence of QDs. As expected, the fluorescence signal of the particles increased with the number of incorporated trioctylphosphine oxide-coated QDs.

2.2.2 Encapsulation of Hydrophobic Molecules

The miniemulsion process is highly suited for the incorporation of all kinds of dyes (e.g., fluorescent dyes which allow the particles to be used as tracer particles). Tronc et al. described the synthesis and characterization of latex particles labeled with a brightly fluorescent yellow dye based on the benzothioxanthene ring structure. The miniemulsion polymerization worked well to yield latex particles of polystyrene, poly(butyl methacrylate), and poly(methyl methacrylate) with high monomer conversion and essentially quantitative dye incorporation [103]. As a different fluorescent dye, a Eu β-ketone complex, which has a long decay time, a large Stokes shift and very narrow emission bands in comparison to other fluorescent dyes, could be used for the incorporation into particles [104]. Highly fluorescent carboxyl and amino functionalized polystyrene particles with defined amounts of the functional groups on the surface were synthesized via miniemulsion polymerization by copolymerizing styrene and acrylic acid or styrene and aminoethyl methacrylate hydrochloride in the presence of the fluorescent dye N-(2,6-diisopropylphenyl)-perylene-3,4dicarboximide [105]. The particles were used as marker particles for uptake into cells. Intracellular localization of amino functionalized particles was confirmed by confocal laser-scanning microscopy (see fluorescent particles in HeLa cell in Figure 2.4(a). It is shown that the uptake increases with increasing surface charges [34].

Lanthanide compounds have, due to their very special electronic structure, a number of extraordinary properties, including their optical, electronic, and magnetic behavior. However, pure lanthanide compounds are usually crystalline and nonductile and have low material performance. Therefore, it is highly profitable for composites with a polymer, thus accessing the potential for solvent processing, the formation of cohesive films, and flexible coating, plus some mechanical advantages, for lanthanide complexes. The addition of hydrophobic neutral, inert inner shell lanthanide complexes such as Gd(III)-(2,2,6,6, tetramethyl-3,5-heptandionate), (tmhd), Eu(III)-(1,1,1,2,2,3,3-heptafluor- 4,6-octandionate) (fod), or Ho(III)- thmd$_3$,

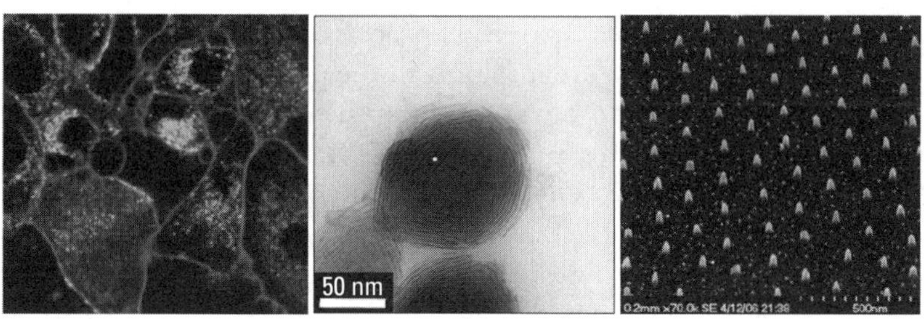

(a) (b) (c)

Figure 2.4 (a) Confocal fluorescent microscopy of HeLa cells after the uptake of amino functionalized fluorescent nanoparticles (green); (b) hybrid nanoparticles with a hydrophobic gadolinium complex and polymethylmethacrylate; and (c) nanopilars from polystyrene nanoparticles filled with hydrophobic Pt complexes after plasma etching.

broadly available as NMR shift reagents, towards ester-containing monomers such as butyl acrylate and subsequent polymerization in miniemulsion droplets leads to the spontaneous formation of highly organized layered nanocomposite particles, as revealed by electron microscopy [see Figure 2.4(b)] and small angle X-ray scattering [94]. Due to the very large coordination numbers of lanthanides and the availability of f-orbitals, those closed-shell complexes however, can bind to additional ligands but also weakly ligating, polarizable components [e.g., Gd $(tmhd)_3$ typically another three], which defines their use as shift reagents. Via those secondary valences, it is possible to mediate compatibility with a second organic phase to allow self-organization and the related formation of nanocomposites. The nanocomposite comprises a lanthanide complex phase and a polymer phase with a lamellar repeat period of about 3.5 nm, rather independent of the system composition. Highly uniform and monodisperse latex particles containing a hydrophobic metal complex as platinum(II)acetylacetonate, indium(III)acetylacetonate, zinc(II)tetramethylheptadionate (Zn(TMHD) 2), zincphthalocyanine, and chromium(III)benzoylacetonate with different loading capacities could be prepared by the miniemulsion polymerization and could be used for a new nanolithography approach. Optimized plasma and annealing procedures generate metal particles with diameters appropriate to their metal content at interparticle distances given by the colloidal size [106]. Serving as etching masks in a reactive-ion etching process, the original order of colloids is transferred to nanopillars and nanoholes with aspect ratios of up to 10; see Figure 2.4(c).

2.2.3 Direct Generation of Polymer Capsules and Hollow Particles

It was also shown that the encapsulation process is not limited on solid materials, but also liquids which are insoluble in the polymeric shell material can be encapsulated in order to obtain nanocapsules. The synthesis of hollow polymeric capsules with sizes ranging from 50 nm to 50 μm has an intense interest in materials science. They offer unique properties as nanoreactors or microreactors and are suitable materials for drug delivery systems. There are generally two approaches to make hollow polymeric particles, which can be differentiated by the presence or absence of a sacrificial core. In the first method, a core particle is used as template and therefore coated with a polymeric shell. The coating of the sacrificial core is obtained by adsorption of a preformed polymer or by polymerization on the surface of the core; subsequently the core is removed by chemical or physical means as dissolution or calcination. The technique of formation of a polymeric shell by adsorption on a sacrificial core comprises the layer-by-layer technique, putting alternatively polyelectrolytes or nanoparticles of opposite charges on a mineral or organic sacrificial core. This method was used for capsule diameters ranging from 70 nm to 10 μm. The thickness of the capsule wall is simply controlled by the number of cycles of the layer-by-layer deposition [107]. Hollow spheres of 200 to 400 nm were also obtained by dissolution or enzymatic degradation of the core of noncovalently connected core-shell micelles [108]. Obviously, the techniques that do not use a sacrificial core are more suitable since they need fewer steps for the synthesis. In this case, the shell formation is driven by self-assembly or by surface tension forces. Thus, capsules can be made by the Ouzo effect [109], the self-assembly of block copolymers [110], the photo-polymerization of a monomer in phospholipid liposomes [111], cross-linking of polymerizable liposomes [112], or neutralization of a polymeric-charged core [113]. The miniemulsion technique was found to be a suitable technique for the synthesis of well-defined nanocapsules because of the high stability of the droplets acting as a nanoreactor.

2.2.4 Encapsulation of Hydrophobic Liquids

For the synthesis, a monomer and an oil are chosen in that way, that both components are mixable in the monomeric state. As soon as polymerization occurs, phase separation takes place. The differences of the hydrophilicities of the interfaces oil/polymer and polymer/water have to be designed so that the formation of nanocapsules is favored [Figure 2.5(a)] [114]. An encapsulation of hydrophobic liquids can also be achieved by using the stabilizer chitosan [see Figure 2.5(b)] which can be transformed to a stable shell by a polyaddition reaction with diepoxides [115]. Several other groups have also reported that hydrophobic compounds could be efficiently encapsulated in thin shells made by free-radical polymerization [116, 117]. An inorganic shell can be formed by adsorbing small silica plates on the surface which are linked by a condensation process [see Figure 2.5(c)] [118].

Torini et al. have described the encapsulation of hydrophobic compounds by interfacial polycondensation in miniemulsion [119]. The process took place in a direct system (o/w) where the stable diisocyanate isophorondiisocyanate (IPDI) was reacted with the 1,6-hexanediol monomer at the interface of the droplets. Scott et al. described an original synthesis of nanocapsules in direct miniemulsion via interfacial free-radical polymerization [120]. A surface-active initiator was used to start an alternating copolymerization between one monomer present in the dispersed oil phase and a monomer present in the aqueous continuous phase.

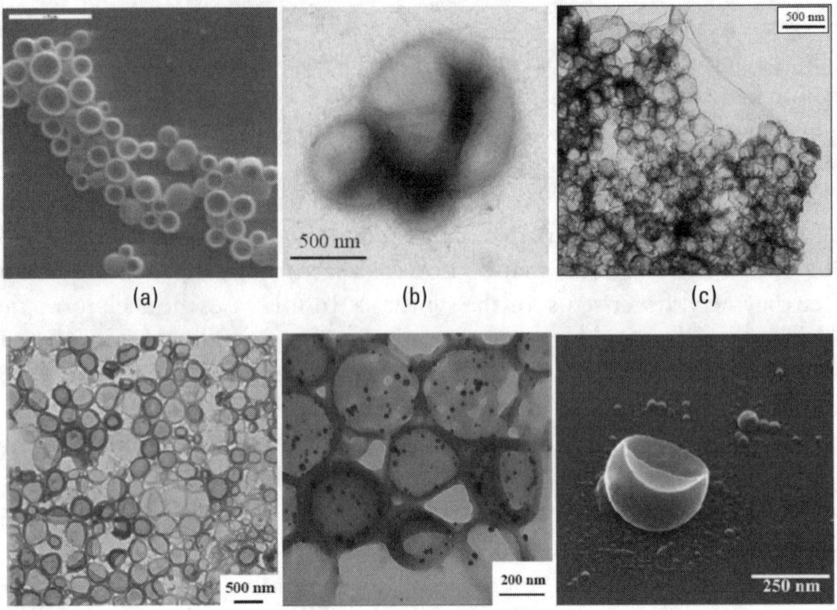

(a) (b) (c) (d) (e) (f)

Figure 2.5 Formation of nanocapsules with: (a) polymer shell containing a hydrophobic liquid; (b) biodegradable chitosan shell containing a hydrophobic liquid; (c) inorganic SiO_2 shell containing a hydrophobic liquid; (d) a polyurethane shell containing a hydrophilic core; (e) silver nanoparticles in polyurethane nanocapsules after reduction of a silver salt; and (f) nanocapsule with encapsulated intact azo component after nanoexplosion.

The wall of the nanocapsules can be either formed that no leakage occurs, or it can be formed as a permeable shell that allows a controlled release (e.g., of a perfume, a medicine, and so forth). As one example, the hydrophobic solid photoinitiator Lucirin TPO was encapsulated within a polymer shell by using the miniemulsion process [121]. A solution of Lucirin TPO in methyl methacrylate (MMA) or butyl acrylate (BA)/MMA mixture was miniemulsified in water followed by a polymerization process in which phase separation of the Lucirin TPO and the formed polymer led to amorphously solidified Lucirin TPO nanoparticles encapsulated by polymer. These nanocapsules were freeze-dried and could be redispersed in acidic monomers, which are applied in polymeric dental adhesives. It is shown by NMR spectroscopy that the shell separates the Lucirin TPO, which is sensitive to degradation in acidic media, from an ambient acidic monomer phase and protects it from fast decomposition. Investigations of the release kinetics of Lucirin TPO from the nanocapsules reveal that the kinetics are strongly dependent on the composition of the surrounding continuous phase.

It could be shown that embedding of intact azoinitiators via radical polymerization of the miniemulsion droplets at low-reaction temperatures has been successfully obtained [122]. In spite of the thermal initiation of the polymerization process with a first azoinitiator decomposing at low temperatures, the embedded second azoinitiators maintain their character so that they can be detonated at a later time at higher temperatures ("nanoexplosion"). The detonation has to be below the glass transition temperature of the polymer. Then the nitrogen gas developed during the thermal treatment of the particles (caused by the decomposition of the encapsulated azoinitiator) build up an overpressure at the inside of the particles and affect a blow-out through which the polymer surface is damaged [see Figure 2.5(f)]. Afterwards the damaged particles collapse. This concept allows a sudden release of materials which are also encapsulated in the particles.

2.2.5 Encapsulation of Hydrophilic Liquids by Interfacial Reaction

A versatile method to obtain functional hollow nanoreactors with a hydrophilic liquid core is to perform an interfacial reaction in miniemulsion [123]. The synthesis of hollow polyurea, polythiourea, and polyurethane nanocapsules was performed by interfacial polycondensation or cross-linking reactions in inverse miniemulsion [see Figure 2.5(d)]. The miniemulsions were built upon emulsification of a solution of amines or alcohols in a polar solvent with cyclohexane as the nonpolar continuous phase. The addition of suitable hydrophobic diisocyanate or diisothiocyanate monomers to the continuous phase allows the polycondensation or the cross-linking reactions to occur at the interface of the droplets. The wall thickness of the capsules can be directly tuned by the quantity of the reactants. The nature of the monomers and the continuous phase are the critical factors for the formation of the hollow capsules, which is explained by the interfacial properties of the system. The resulting polymer nanocapsules could be subsequently dispersed in water. The capsules were found to be spherical when formamide was used as the liquid core, whereas elongated capsules were obtained with water. Finally, these hollow nanoreactors were used as a model system for the preparation of silver nanoparticles by reducing silver nitrate solutions encapsulated by the polyurea shell [see Figure 2.5(e)]. These syntheses are the first that allow the encapsulation of hydrophilic compounds in miniemulsion in a hollow structure.

Nanocapsules with polymeric shells of polyurethane, polyurea, and crosslinked dextran containing the hydrophilic contrast agents Magnevist and Gadovist are synthesized via the same process [124]. Due to the porosity of the polymeric shell, exchange of water molecules through the capsule walls can be ensured and no significant compromises in the T_1 relaxitivity of the contrast compound is observed for magnetic resonance imaging. This clearly indicates the potential use of the contrast-agent-filled nanocapsules as new magnetic resonance imaging (MRI) contrast agents. Aqueous-core capsules with uniform polymeric shells and diameters ranging from 0.2 to 5 μm could also be prepared by polymerizing the interfaces of inverse emulsion microspheres [125]. The free-radical polymerization at the interface of water-in-oil microspheres was performed as an alternating copolymerization of hydrophobic maleate esters and hydrophilic polyhydroxy vinyl ethers, in a manner analogous to classical interfacial polycondensations. In these polymerizations, the kinetics, shell thickness, and release characteristics of the resulting aqueous-core capsules are set by the diffusion-limited alternating reaction of the oil-soluble maleate esters and water-soluble vinyl ethers.

2.2.6 Encapsulation of Hydrophilic Components by Nanoprecipitation

The modified nanoprecipitation of polymers onto stable nanodroplets has been successfully applied to prepare well-defined nanocapsules whose hydrophilic core is composing of an antiseptic agent (i.e., chlorhexidine digluconate aqueous solution) [126]. The stable nanodroplets were obtained by inverse miniemulsions with an aqueous antiseptic solution dispersed in an organic medium of solvent/nonsolvent mixture containing an oil-soluble surfactant and the polymer for the shell formation. The change of gradient of the solvent/nonsolvent mixture of dichloromethane/cyclohexane, obtained by heating at 50°C, led to the precipitation of the polymer in the organic continuous phase and deposition onto the large interface of the miniemulsion aqueous droplets. The monodisperse polymer nanocapsules with the size range of 240 to 80 nm were achieved as a function of the amount of surfactant. Using various polymer contents, molecular weights and types, an encapsulation efficiency of 20% to 100% was obtained. The nanocapsules could be easily transferred into water as continuous phase resulting in nanocapsules containing an aqueous core with the antiseptic agent. The encapsulated amount of the antiseptic angent was evaluated and indicated the durability of the nanocapsule's wall.

2.3 Crystallization in Miniemulsion Droplets

Due to the small size of the droplets, the dynamic crystallization and melting process in small, stable, and narrowly distributed nanodroplets of miniemulsions is strongly influenced. Both regular and inverse systems were examined, characterizing in the one case on the crystallization of n-alkanes [127, 128], the crystallization of ice in the other [127]. It was shown for both cases that the temperature of crystallization in such droplets is significantly decreased (or the required undercooling is increased), as compared to the bulk material. This can be attributed to a very effective suppression of heterogeneous nucleation. This means that a crystallization does not occur in the binodal region of the phase diagram, but much deeper below the spinodal line which increases the region of metastability by another 25K. This is important for a whole variety of technical applications.

Interestingly, the confinement is droplets and also influences the crystal morphology and crystal structure, as detected by X-ray analysis. The more plastic hexadecane crystals show a distortion of the crystal structure by the adaptation to the close-to-spherical shape. A very different behavior was detected for odd and even alkanes. In even alkanes, the confinement in small droplets changes the crystal structure from a triclinic (in bulk) to an orthorhombic structure which is attributed to finite size effects inside the droplets. An intermediate metastable rotator phase is of less relevance for the miniemulsion droplets than in the bulk. For odd alkanes, only a strong temperature shift compared to the bulk system, but no structure change is observed, as both in bulk and in miniemulsion droplets, a triclinic structure is formed.

Water, on the other hand, shows the same hexagonal structure, but the relative peak intensities heavily change, speaking for a very flat shape of ice nanocrystal. At the same time, it is that more than one ice crystal nucleate in each nanodroplet, making the crystalline nanodroplet superstructure potentially looking like a pile of pancakes. The nanocrystal size in case of ice is increasing with decreasing droplet size, which can be due to heat flow effects, a decreased nucleation rate, or even a better packing of the nanocrystals in smaller droplets. Crystallization of the polymer poly(ethylene oxide) in narrowly distributed nanodroplets also show exclusively homogeneous nucleation resulting in stable crystallized droplets [129]. Inside each droplet, 4–5 lamellae are formed which are not interlamellar connected and just loosely layered. During the crystallization, in each droplet, only one nucleus is present at the same time. At the large supercooling only about 60% of the chains are crystallized indicating a high imperfection of the superstructure.

However, a rearrangement occurs at increasing temperature. Whereas in dispersion the structure is stable, upon drying the lamellae slid apart from each other and arrange in a highly ordered, Zenon-structural way. The smallest units (the "cap" of the droplet) indeed consists of down to one polymer chain. Therefore, the crystallization of polymers confined in miniemulsion can be also used to obtain single-chain single crystals.

The crystallization of the miniemulsions of two dyes (oil blue and oil red) with primary droplet diameters of 120 nm results in single crystalline nanofibers of high quality, uniformity, and chromatic definition [130]. As these dyes are absolutely insoluble in the continuous phase, the observed growth of the crystals must have proceeded via controlled aggregation and mesoscale transformation of colloidal intermediates. This is regarded as a model case for this nonclassical crystallization process.

The fact that the crystals of those dyes are pleiochromic reveals additional information about the aggregation process. Dye absorption and therefore the maximum dipole moment are oriented perpendicular to the growth direction, which gives strong indications that the controlled aggregation is mediated not by dipole fields (as usually speculated), but by polarization forces. As these, for polar crystals, add up coherently in a highly anisotropic fashion, van der Waals attraction in certain direction can obviously become very strong, much stronger than the ionic and steric stabilizers which have kept the original miniemulsion stable. It was called a "super van der Waals" force which is highly directional and breaks the radial symmetry of the DLVO potential. As a result, highly selective and spatially controlled aggregation takes place, which is the prerequisite of morphosynthetical control of the crystal habitus. In addition, the existence of such forces also explain why industrially optimized procedures to make dye or drug nanocrystals succeed or fail from system to system in a way nonpredictable from the properties of the molecule, alone. In this picture, stable nanocrystals can only be made from crystal structures which do not show coherent addition of molecular polarizabilities.

Liquid crystal nanoparticles using a low molecular weight liquid crystal were also prepared by the miniemulsion approach with a droplet size between 180 to 630 nm [131]. DSC measurements reveal a large shift in the nematic-isotropic phase transition temperature. Further investigations on liquid crystal droplets are carried out using AFM measurements showing that these particles have an order of the liquid crystal molecules within the droplets. Light scattering measurements yield the temperature dependence of the anisotropy and the temporal stability of the droplets. An effect of the director fluctuations within the droplets on dynamic depolarized scattering was found.

2.4 Conclusion

Dispersions of liquid matter in stable nanodroplets as nanoreactors in a so-called "miniemulsion" open new possibilities for the synthesis of nanoparticles and nanocapsules. The entire range of polymerization reactions from radical, anionic, cationic, enzymatic polymerization to polyaddition and polycondensation, metal-catalyzed, and oxidative polymerization can be performed in order to obtain different polymeric particles. Inorganic particles and liquids can be encapsulated by a subsequent second miniemulsion process into a polymer shell in order to avoid leakage. This protects the interior against external influences, but may also protect the environment (e.g., the human body against toxic materials). A permeable shell allows the controlled slow release of substances into the environment; fast releases can also be obtained by nanoexplosions inside the nanocapsules. The strength of miniemulsion is that polymeric nanoparticles and nanocapsules can be produced consisting of polymers or polymer structures including functional surfaces and encapsulated materials which are not accessible by other types of heterophase polymerization. In my opinion, the field of miniemulsion is still on the rise since there are many possibilities for the design of new particles covering many applications in materials science for life sciences and medicine.

References

[1] Couvreur, P.; Barratt, G.; Fattal, E.; Legrand, P., *Crit. Rev. in Therap. Drug Carr. Syst.* 2002, 19, 99.

[2] Antonietti, M.; Förster, *S. Adv. Mater.* 2003, 15, 1323.

[3] Sukhorukov, G. B.; Fery, A.; Brumen, M.; Möhwald, H., *Phys. Chem. Chem. Phys.* 2004, 6, 4078.

[4] Dawkins, J. V., In *Aqueous Suspension Polymerization in Comprehensive Polymer Science,* Allen; Bevington, Eds. Pergamon Press: Oxford, 1989; pp 231-241.

[5] Ugelstad, J.; El-Aasser, M. S.; Vanderhoff, J. W., *J. Polym. Sci., Polym. Lett. Ed.,* 1973, 11, 503-513.

[6] Blythe, P. J.; Sudol, E. D.; El-Aasser, M. S., *Macromolecular Symposia 2000,* 150, 179–186.

[7] Schork, F. J.; Poehlein, G. W.; Wang, S.; Reimers, J.; Rodrigues, J.; Samer, C., "Colloids Surf." *A: Physicochem. Eng. Asp.* 1999, 153, 39-45.

[8] Landfester, K. *Macromol. Rapid Comm.* 2001, 22, 896-936.

[9] Lowe, K. C. "Art. Cells, Blood Subs., and Immob." *Biotech.* 2000, 28, (1), 25-38.

[10] Back, A. J.; Schork, F. J., *Journal of Applied Polymer Science,* 2007 103, (2), 819-833.

[11] Landfester, K.; Rothe, R.; Antonietti, M. *Macromolecules 2002,* 35, (5), 1658-1662.

[12] Landfester, K.; Antonnietti, M., *Macromol. Rapid Com.*, 2000, 21, pp. 820-824.

[13] Delgado, J.; EL-Aasser, M. S.; Silibi, C. A.; Vanderhoff, J. W., *J. Polym. Sci., Polym. Chem. Ed.* 1990, 28, 777-794.

[14] Wang, S.; Schork, F. J., *J. Appl. Polym. Sci.* 1994, 54, 2157-2164.

[15] Wu, X. Q.; Schork, F., *J. Industrial & Engineering Chemistry Research* 2000, 39, (8), 2855-2865.

[16] Tronc, F.; Winnik, M. A.; Kaul, B. L.; Graciet, J. C., *Journal of Polymer Science Part a-Polymer Chemistry,* 2004, 42, (8), 1999-2009.

[17] Landfester, K.; Willert, M.; Antonietti, M., *Macromolecules* 2000, 33, 2370-2376.

[18] Willert, M., *Inverse Miniemulsions.* Ph.D. thesis, Universität Potsdam, 2001.

[19] Blagodatskikh, I.; Tikhonov, V.; Ivanova, E.; Landfester, K.; Khokhlov, A. *Macromolecular Rapid Communications,* 2006, 27, (22), 1900-1905.

[20] Rodriguez, V. S.; El-Aasser, M. S.; Asua, J. M.; Silibi, C. A., *J. Polym. Sci., Polym. Chem. Ed.,* 1989, 27, 3659-3671.

[21] Huang, H.; Zhang, H.; Li, J.; Cheng, S.; Hu, F.; Tan, B., *J. Appl. Polym. Sci.,* 1998, 68, 2029-2039.

[22] Delgado, J.; El-Aasser, M. S.; Vanderhoff, J. W., *J. Polym. Sci., Polym. Chem. Ed.,*1986, 24, 861-874.

[23] Jovanovic, R.; Dube, M. A., *Macromolecular Chemistry and Physics,* 2004, 205, (7), 958–965.

[24] Li, D.; Sudol, E. D.; El-Aasser, M. S., *Journal of Applied Polymer Science,* 2006, 102, (5), 4616-4622.

[25] Li, D. H.; Sudol, E. D.; El-Aasser, M. S., *Journal of Applied Polymer Science,* 2006, 101, (4), 2304-2312.

[26] Jovanovic, R.; Ouzineb, K.; McKenna, T. F.; Dube, M. A., *Macromolecular Symposia,* 2004, 206, 43-56.

[27] Reimers, J.; Schork, F., *J. Polym. Reaction Eng.,* 1996, 4, 135-152.

[28] Pich, A.; Datta, S.; Musyanovych, A.; Adler, H. J. P.; Engelbrecht, L. *Polymer,* 2005, 46, (4), 1323-1330.

[29] Landfester, K.; Pawelzik, U.; Antonietti, M., *Polymer,* 2005, 46, 9892-9898.

[30] Chern, C. S.; Sheu, J. C., *Polymer,* 2001, 42, (6), 2349-2357.

[31] Chern, C. S.; Sheu, J. C., *Journal of Polymer Science Part a-Polymer Chemistry,* 2000, 38, (17), 3188-3199.

[32] Willert, M.; Landfester, K., *Macromol. Chem. Phys.,* 2002, 203, 825-836.

[33] Luo, Y. W.; Schork, F. J., *Journal of Polymer Science Part a-Polymer Chemistry,* 2001, 39, (16), 2696-2709.

[34] Lorenz, M. R.; Holzapfel, V.; Musyanovych, A.; Nothelfer, K.; Walther, P.; Frank, H.; Landfester, K.; Schrezenmeier, H.; Mailander, V., *Biomaterials,* 2006, 27, (14), 28202828.

[35] Herrera, V.; Pirri, R.; Asua, J. M.; Leiza, J. R., *Journal of Polymer Science Part a-Polymer Chemistry,* 2007, 45, (12), 2484-2493.

[36] Koenig, A.; Ziener, U.; Schaz, A.; Landfester, K., *Macromolecular Chemistry and Physics,* 2007, 208, (2), 155-163.

[37] Landfester, K.; Montenegro, R.; Scherf, U.; Güntner, R.; Asawapirom, U.; Patil, S.; Kietzke, T.; Neher, D., *Adv. Mater.,* 2002, 14, (9), 651-655.

[38] Asawapirom, U.; Bulut, F.; Farrell, T.; Gadermaier, C.; Gamerith, S.; Güntner, R.; Kietzke, T.; Patil, S.; Piok, T.; Montenegro, R.; Stiller, B.; Tiersch, B.; Landfester, K.; List, E. J. W.; Neher, D.; Torres, C. S.; Scherf, U., *Macromol. Symp.,* 2004, 212, 83–91.

[39] Kietzke, T.; Neher, D.; Landfester, K.; Montenegro, R.; Güntner, R.; Scherf, U., *Nature Materials,* 2003, 2, (6), 408-412.

[40] Kietzke, T.; Neher, D.; Kumke, M.; Montenegro, R.; Landfester, K.; Scherf, U., *Macromolecules,* 2004, 37, 4882-4890.

[41] Kietzke, T.; Neher, D.; Kumke, M.; Ghazy, O.; Ziener, U.; Landfester, K., *Small,* 2007, 3, (6), 1041-1048.

[42] Matyjaszewski, K.; Gaynor, S. G., "Free radical polymerization." In *Applied Polymer Science 21th Century,* Craver, C. D.; Carraher, C. E., Eds. Elsevier Science Ltd.: Oxford, 2000; pp 929-977.

[43] MacLeod, P. J.; Barber, R.; Odell, P. G.; Keoshkerian, B.; Georges, M. K., *Macromolecular Symposia,* 2000, 155, 31-38.

[44] Prodpran, T.; Dimonie, V. L.; Sudol, E. D.; El-Aasser, M. S., *Macromol. Symp.,* 2000, 155, 1-14.

[45] Cunningham, M. F.; Tortosa, K.; Ma, J. W.; McAuley, K. B.; Keoshkerian, B.; Georges, M. K., *Macromolecular Symposia,* 2002, 182, 273-282.

[46] Enright, T. E.; Cunningham, M. F.; Keoshkerian, B., *Macromolecular Rapid Communications,* 2005, 26, (4), 221-225.

[47] Nicolas, J.; Charleux, B.; Guerret, O.; Magnet, S., *Macromolecules,* 2004, 37, (12), 4453–4463.

[48] Farcet, C.; Nicolas, J.; Charleux, B., *Journal of Polymer Science Part a-Polymer Chemistry,* 2002, 40, (24), 4410-4420.

[49] Farcet, C.; Charleux, B., *Macromolecular Symposia,* 2002, 182, 249-260.

[50] Delaittre, G.; Nicolas, J.; Lefay, C.; Save, M.; Charleux, B., *Chemical Communications,* 2005, (5), 614-616.

[51] Cunningham, M. F.; Ng, D. C. T.; Milton, S. G.; Keoshkerian, B., *Journal of Polymer Science Part a-Polymer Chemistry,* 2006, 44, (1), 232-242.

[52] de Brouwer, H.; Tsavalas, J. G.; Schork, F. J.; Monteiro, M., *J. Macromolecules,* 2000, 33, (25), 9239-9246.

[53] Vosloo, J. J.; De Wet-Roos, D.; Tonge, M. P.; Sanderson, R. D., *Macromolecules,* 2002, 35, (13), 4894-4902.

[54] Shim, S. E.; Lee, H.; Choe, S., *Macromolecules,* 2004, 37, (15), 5565-5571.

[55] Simms, R. W.; Davis, T. P.; Cunningham, M. F., *Macromolecular Rapid Communications,* 2005, 26, (8), 592-596.

[56] Bowes, A.; McLeary, J. B.; Sanderson, R. D., *Journal of Polymer Science Part a-Polymer Chemistry,* 2007, 45, (4), 588-604.

[57] Smulders, W. W.; Jones, C. W.; Schork, F. J., *Macromolecules,* 2004, 37, (25), 9345-9354.

[58] Smulders, W. W.; Jones, C. W.; Schork, F. J. *Aiche Journal,* 2005, 51, (3), 1009-1021.

[59] Russum, J. P.; Jones, C. W.; Schork, F. *J. Industrial & Engineering Chemistry Research,* 2005, 44, (8), 2484-2493.

[60] Matyaszewski, K.; Qiu, J.; Tsarevsky, N. V.; Charleux, B., *J. Polym. Sci., Polym. Chem. Ed.,* 2000, 38, 4724-4734.

[61] Qiu, J.; Pintauer, T.; Gaynor, S. G.; Matyjaszewski, K.; Charleux, B.; Vairon, J. P., *Macromolecules,* 2000, 33, (20), 7310-7320.

[62] Li, M.; Matyjaszewski, K., *Journal of Polymer Science Part a-Polymer Chemistry,* 2003, 41, (22), 3606-3614.

[63] Min, K.; Gao, H. F.; Matyjaszewski, K., *Journal of the American Chemical Society,* 2005, 127, (11), 3825-3830.

[64] Min, K.; Jakubowski, W.; Matyjaszewski, K., *Macromolecular Rapid Communications,* 2006, 27, (8), 594-598.

[65] Li, M.; Jahed, N. M.; Min, K.; Matyjaszewski, K., *Macromolecules,* 2004, 37, (7), 24342441.

[66] Oh, J. K.; Siegwart, D. J.; Lee, H. I.; Sherwood, G.; Peteanu, L.; Hollinger, J. O.; Kataoka, K.; Matyjaszewski, K., *Journal of the American Chemical Society,* 2007, 129, (18), 5939-5945.

[67] Maitre, C.; Ganachaud, F.; Ferreira, O.; Lutz, J. F.; Paintoux, Y.; Hemery, P., *Macromolecules,* 2000, 33, (21), 7730-7736.

[68] Yactine, B.; Ganachaud, F.; Senhaji, O.; Boutevin, B., *Macromolecules,* 2005, 38, (6), 2230-2236.

[69] Ivanenko, C.; Maitre, C.; Ganachaud, F.; Hemery, P., *E-Polymers,* 2003.

[70] Rehor, A.; Tirelli, N.; Hubbell, J. A., *Macromolecules,* 2002, 35, (23), 8688-8693.

[71] Weiss, C. K.; Ziener, U.; Landfester, K., *Macromolecules,* 2007, 40, (4), 928-938.

[72] Weiss, C. W.; Lorenz, M. R.; Landfester, K.; Mailänder, V., *Macromol. Biosci.,* 2007, 7, (7), 883-896.

[73] Crespy, D.; Landfester, K., *Macromolecules,* 2005, 38, 6882-6887.

[74] Cauvin, S.; Ganachaud, F., *Macromolecular Symposia,* 2004, 215, 179-189.

[75] Cauvin, S.; Ganachaud, F.; Moreau, M.; Hemery, P., *Chemical Communications,* 2005, (21), 2713-2715.

[76] Kostjuk, S. V.; Ganachaud, F., *Macromolecules,* 2006, 39, (9), 3110-3113.

[77] Taden, A.; Antonietti, M.; Landfester, K., *Macromol. Rapid Comm.,* 2003, 24, 512-516.

[78] Marie, E.; Rothe, R.; Antonietti, M.; Landfester, K., *Macromolecules,* 2003, 36, (11), 3967-3973.

[79] Mecking, S., *Colloid and Polymer Science,* 2007, 285, (6), 605-619.

[80] Bauers, F. M.; Mecking, S., *Angewandte Chemie-International Edition,* 2001, 40, (16), 3020-3022.

[81] Bauers, F. M.; Chowdhry, M. M.; Mecking, S., *Macromolecules,* 2003, 36, (18), 67116715.

[82] Monteil, V.; Bastero, A.; Mecking, S., *Macromolecules,* 2005, 38, (13), 5393-5399.

[83] Held, A.; Kolb, L.; Zuideveld, M. A.; Thomann, R.; Mecking, S.; Schmid, M.; Pietruschka, R.; Lindner, E.; Khanfar, M.; Sunjuk, M., *Macromolecules,* 2002, 35, (9), 3342-3347.

[84] Klapper, M.; Jang, Y. J.; Bieber, K.; Nemnich, T.; Nenov, N.; Mullen, K., *Macromolecular Symposia,* 2004, 213, 131-145.

[85] Chemtob, A.; Gilbert, R. G., *Macromolecules,* 2005, 38, (16), 6796-6805.

[86] Quemener, D.; Chemtob, A.; Heroguez, V.; Gnanou, Y., *Polymer,* 2005, 46, (4), 10671075

[87] Mingotaud, A. F.; Kramer, M.; Mingotaud, C., *Journal of Molecular Catalysis a-Chemical,* 2007, 263, (1-2), 39-47..

[88] Landfester, K.; Tiarks, F.; Hentze, H.-P.; Antonietti, M., *Macromol. Chem. Phys.,* 2000, 201, 1-5.

[89] Tiarks, F.; Landfester, K.; Antonietti, M., *J. Polym. Sci., Polym. Chem. Ed.,* 2001, 39, (14), 2520-2524.

[90] Barrere, M.; Landfester, K., *Macromolecules,* 2003, 36, (14), 5119-5125.

[91] Barrere, M.; Landfester, K., *Polymer,* 2003, 44, (10), 2833-2841.

[92] Musyanovych, A.; Mailaender, V.; Landfester, K., *Biomacromolecules,* 2005, 6, 18241828.

[93] Bechthold, N.; Tiarks, F.; Willert, M.; Landfester, K.; Antonietti, M., *Macromol. Symp.,* 2000, 151, 549-555.

[94] Ramirez, L. P.; Antonietti, M.; Landfester, K., *Macromolecular Chemistry and Physics,* 2006, 207, (2), 160-165.

[95] Tiarks, F.; Landfester, K.; Antonietti, M., *Macromol. Chem. Phys.,* 2001, 202, 51-60.

[96] Ha, M. L. P.; Grady, B. P.; Lolli, G.; Resasco, D. E.; Ford, W. T., *Macromolecular Chemistry and Physics,* 2007, 208, (5), 446-456.

[97] Steiert, N.; Landfester, K., *Macromol. Mater. Eng.,* 2007.

[98] Hoffmann, D.; Landfester, K.; Antonietti, M., *Magnetohydrodynamics,* 2001, 37, (3), 217221.

[99] Ramirez, L. P.; Landfester, K., *Macromol. Chem. Phys.,* 2003, 204, 22-31.

[100] Landfester, K.; Ramirez, L., *Journal of Physics: Condensed Matter,* 2003, 15, (15), S13451362.

[101] Holzapfel, V.; Lorenz, M.; Weiss, C. K.; Schrezenmeier, H.; Landfester, K.; Mailander, V., *Journal of Physics-Condensed Matter,* 2006, 18, (38), S2581-S2594.

[102] Joumaa, N.; Lansalot, M.; Theretz, A.; Elaissari, A., *Langmuir,* 2006, 22, (4), 1810-1816

[103] Tronc, F.; Li, M.; Lu, J. P.; Winnik, M. A.; Kaul, B. L.; Graciet, J. C., *Journal of Polymer Science Part a-Polymer Chemistry,* 2003, 41, (6), 766-778.

[104] Ando, K.; Kawaguchi, H., *Journal of Colloid and Interface Science,* 2005, 285, (2), 619626.

[105] Holzapfel, V.; Musyanovych, A.; Landfester, K.; Lorenz, M. R.; Mailander, V., *Macromolecular Chemistry and Physics,* 2005, 206, (24), 2440-2449.

[106] Manzke, A.; Pfahler, C.; Dubbers, O.; Plettl, A.; Ziemann, P.; Crespy, D.; Schreiber, E.; Ziener, U.; Landfester, K., *Advanced Materials,* 2007, 19, (10), 1337-+.

[107] Sukhorukov, G. B.; Donath, E.; Lichtenfeld, H.; Knippel, E.; Knippel, M.; Budde, A.; Möhwald, H., *Colloids Surf. A: Physicochem. Eng. Asp.,* 1998, 137, 253-266.

[108] Duan, H.; Chen, D.; Jiang, M.; W. Gan; Li, S.; Wang, M.; Gong, J. J., *J. Am. Chem. Soc.,* 2001, 123, 12097-12098.

[109] Ganachaud, F.; Katz, J. L., *Chemphyschem,* 2005, 6, (2), 209-216.

[110] Wong, M. S.; Cha, J. N.; Choi, K.; Deming, T. J.; Stucky, D., *Nano Lett.,* 2002, 2, pp. 583-587.

[111] Lestage, D. J.; Urban, M., *Langmuir,* 2005, 21, 4266-4267.

[112] Liu, S.; O'Brien, D. F., *J. Am. Chem. Soc.,* 2002, 124, 6037-6042.

[113] Pavlyuchenko, V. N.; Sorochinskaya, O. V.; Irvanche, S. S.; Klubin, V. V.; Kreichman, G. S.; Budto, V. P.; Skrifvars, M.; Halme, E.; Koskinen, J., *J. Polym. Sci., Part A: Polym. Chem.,* 2001, 39, 1435-1449.

[114] Tiarks, F.; Landfester, K.; Antonietti, M., *Langmuir,* 2001, 17, (3), 908-917.

[115] Marie, E.; Landfester, K.; Antonietti, M., *Biomacromolecules,* 2002, 3, (3), 475-481.

[116] Rajot, I.; Bone, S.; Graillat, C.; Hamaide, T., *Macromolecules,* 2003, 36, (20), 7484-7490.

[117] Luo, Y.; X., Z., *J. Polym Sci., Part A : Polym. Chem.,* 2004, 42, 2145-2154.

[118] Putlitz, B. z.; Landfester, K.; Fischer, H.; Antonietti, M., *Adv. Mater.,* 2001, 13, (7), 500503.

[119] Torini, L.; Argillier, J. F.; Zydowicz, N., *Macromolecules,* 2005, 38, (8), 3225-3236.

[120] Scott, C.; Wu, D.; Ho, C. C.; Co, C. C., *Journal of the American Chemical Society,* 2005, 127, (12), 4160-4161.

[121] Volz, M.; Ziener, U.; Salz, U.; Zimmermann, J.; Landfester, K., *Colloid and Polymer Science,* 2007, 285, (6), 687-692.

[122] Volz, M.; Walther, P.; Ziener, U.; Landfester, K., *Macromol. Mater. Eng.,* 2007.

[123] Crespy, D.; Stark, M.; Hoffmann-Richter, C.; Ziener, U.; Landfester, K., *Macromolecules,* 2007, 40, (9), 3122-3135.

[124] Jagielski, N.; Sharma, S.; Hombach, V.; Mailänder, V.; Rasche, V.; Landfester, K., *Macromol. Chem. Phys.,* 2007.

[125] Wu, D.; Scott, C.; Ho, C. C.; Co, C. C., *Macromolecules,* 2006, 39, (17), 5848-5853.

[126] Paiphansiri, U.; Tangboriboonrat, P.; Landfester, K., *Macromolecular Bioscience,* 2006, 6, (1), 33-40.

[127] Montenegro, R.; Antonietti, M.; Mastai, Y.; Landfester, K., *J. Phys. Chem.*, 2003, 107, (21), 5088-5094.

[128] Montenegro, R.; Landfester, K., *Langmuir,* 2003, 19, (15), 5996-6003.

[129] Taden, A.; Landfester, K., *Macromolecules,* 2003, 36, (11), 4037-4041.

[130] Taden, A.; Landfester, K.; Antonietti, M., *Langmuir,* 2004, 20, (3), 957-961

[131] Tongcher, O.; Sigel, R.; Landfester, K., *Langmuir,* 2006, 22, (10), 4504-4511.

3

Transport Phenomena and Chemical Reactions in Nanoscale Surfactant Networks

Aldo Jesorka, Owe Orwar, Zoran Konkoli, Brigitte Bauer, and Ludvig Lizana

3.1 Introduction

Biological cells have evolved over time into sophisticated, highly structured chemical factories. They feature spatially confined mechanisms for transport, storage, release, and mixing of reactants, catalysts, and products down to single molecule quantities (see Figure 3.1) (Alberts, Johnson, et al., 2002). On the cellular level, nanosized biochemical reactors are abundant, with important functions in energy conversion, cell signaling, processing of genetic material, and synthesis of a great variety of essential functional materials. So far, artificially created small-scale reactor systems have never reached this structural complexity and diversity. However, pathways to new technologies begin to open up, with soft materials being used to create artificial environments with cell-like properties. They make possible the construction of reactor assemblies that can be dynamically controlled, modified, and restructured in order to operate with very small numbers of molecules. The necessary initiation and control of chemical processes at this scale in artificial

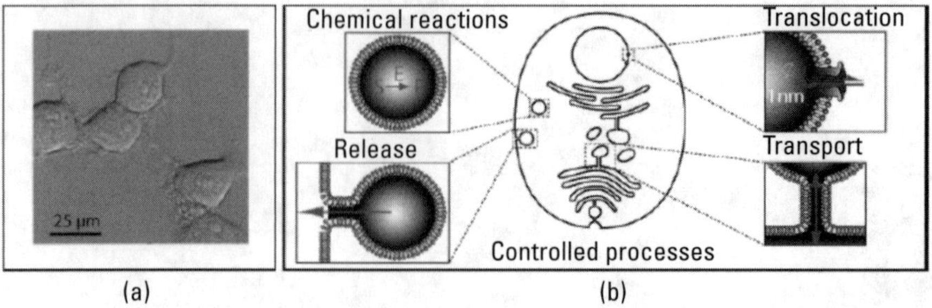

Figure 3.1 Inspiration from biology, where solutions for handling single molecules and extremely small volumes already exist. (a) Micrograph of adherent eukaryotic cells (NG 108-15), (b) schematic of the endomembrane system (formed by endoplasmic reticulum Golgi, etc.), in which controlled reactions, controlled release (and uptake) of material, controlled translocation, and controlled transport are taking place.

environments can be achieved by combining biologically inspired materials and principles with man-made, engineered materials (Wu and Payne, 2004).

This chapter highlights recent advances in soft-matter nanotechnology, largely based on phospholipid biomembrane assemblies. The information found here should provide biomedical researchers and professionals in the biophysical sciences with state-of-the-art material required to understand and evaluate the opportunities and challenges of soft-matter reactor systems on the nano- and microscale. The first section is dedicated to the fabrication and properties of vesicle-nanotube-networks and their constituents, the second section is dedicated to diffusion and transport mechanisms in such systems, and the third section introduces aspects of chemical reactions in the networks, largely based on application examples.

While today's microreaction and microscale transport devices (microfluidics, microelectrofluidics) are established technological areas (Brivio, Verboom, et al., 2006) (Weibel and Whitesides, 2006) (Jayaraj, Kang, et al., 2007), engineering of chemical reactor systems of nanometer dimensions and nanobiotechnological applications are still at an early stage (Oberholzer and Luisi, 2002) (Bolinger, Stamou, et al., 2004). Researchers in chemistry, physics, and materials science have collected extensive knowledge about nanoscale structural principles found in various instances in the biological world (Clark, Singer et al., 2004) (Pitchiaya and Krishnan, 2006) (Mavroidis, Dubey, et al., 2004) (Koh and Moon, 2006). The driving force for such investigations lays in the desire to gain deeper knowledge of, for example, rates and mechanisms of biochemical reactions in confined environments, in the need for novel applications and technologies such as ultra-small-scale analytical devices or in the search for unconventional computing paradigms (Zauner, 2005). Accordingly, many new observation and nanoengineering techniques have emerged recently, and the advantages of nanosize devices, such as low energy/material consumption, high efficiency, and specific functionality, are apparent. Now the field is driving towards the development of methods to bring such nanoscale (bio)-technologies into service.

Self-organization and molecular self-assembly (Huie, 2003), chemical information processing (e.g., by means of DNA) (Banzhaf, Dittrich, et al., 1996), the operation principles of molecular motors (Mallik and Gross, 2004), or the favorable material properties of a soft-matter architecture as found in biomembranes (Tien, Salamon, et al., 1991) (Kinoshita, 1995) are all examples for promising leads taken directly from nature. Specifically, in order to construct structured biochemical reactors with, at least in part, nanoscale dimensions, the use of construction principles and material choices found in the living world is one of the most promising approaches.

A soft-matter device technology based on biomembranes that is easily constructed and readily transformed into complex functional units of microreactors and interconnecting flexible nanochannels is the essence of this chapter. Surfactant nanotube-vesicle networks represent some of the smallest devices known to date for performing controlled chemical operations in a completely biocompatible environment. Newly developed means for compartmentalization and transport of materials between containers, as well as advancements in initiation and control of chemical reactions in such systems have opened pathways to new devices with applications down to the single-molecule level (Karlsson, Karlsson, et al., 2006).

3.2 Construction, Shape Transformations, and Structural Modifications of Phospholipid Nanotube-Vesicle Networks

3.2.1 Phospholipid Membranes and Vesicles

Although the number of self-assembling soft materials is rapidly growing (Hamley, 2005) (Kato, Mizoshita, et al., 2006) (Ghoroghchian, Frail, et al., 2005), the most widely applied bilayer-forming material in the biosciences is phospholipids. They are successfully utilized in the assembly of supported membranes (Tanaka and Sackmann, 2005) (Zana, 2005), membrane-enclosed pico-to-femtoliter volumes (vesicles) (Houser, 1993) (Taylor, Davidson, et al., 2005) (Luisi and Walde, 1999), (Segota and Tezak, 2006) and complexly structured networks of vesicles interconnected by tubular structures (Karlsson, Karlsson, et al., 2001).

Biological cells are surrounded by bilayer membranes, which are mainly composed of phospholipids, proteins, and carbohydrates arranged in a fluidic mosaic structure. From the perspective of a biomaterials scientist, lipid bilayer membranes have several unique and attractive properties. They consist of tightly packed, amphiphilic phospholipids molecules, held together by nonconvalent interactions. These molecules are neatly arranged in a double layer, with their polar, hydrophilic phosphate heads facing outwards, and their nonpolar, hydrophobic fatty acid tails facing each other. This essentially hydrophobic bilayer acts as a barrier to all ionic and many uncharged species, effectively separating the two opposite sides of the membrane from each other. Different kinds of membranes can contain a variety of differently structured phospholipids as well as other constituents (e.g., cholesterol in animal cell membranes), affecting both strength and flexibility of the bilayer.

Unlike solid materials or rigid macromolecular structures, supramolecular bilayers possess extraordinary mechanical properties as well as features of a two-dimensional fluid. The mechanical strength of a fluid-state lipid membrane is comparable to a steel sheet of the same thickness, yet it is capable of undergoing complex shape transitions (Seifert, 1997). A theoretical framework for describing the elastic properties of vesicles has been developed and constantly improved (Evans and Skalak, 1980) (Dobereiner, Evans, et al., 1997). The most important macroscopic property of lipid bilayers is the surface bending elasticity, which is closely related to vesicle shape, the chemical properties of individual phospholipids in the membrane, and the nature of structural phases and their transitions (Evans and Needham, 1987) (Needham and Zhelev, 1996).

3.2.2 Self-Assembly of Vesicular Systems

Most phospholipids spontaneously self-aggregate to ~5-nm-thick bilayer membranes when suspended in an aqueous solution. Depending on the assembly conditions and the chemical structure of the phospholipids, planar bilayers, spherical vesicles (liposomes), or more complex aggregates such as cubosomes can be obtained (Barauskas, Johnsson, et al., 2005). Membranes assembled through this self-association mechanism can, under certain conditions, completely encapsulate volumes of the medium they were suspended in, forming nearly spherical, flexible compartments (liposomes) with an interior volume that is isolated from the external medium (Figure 3.2).

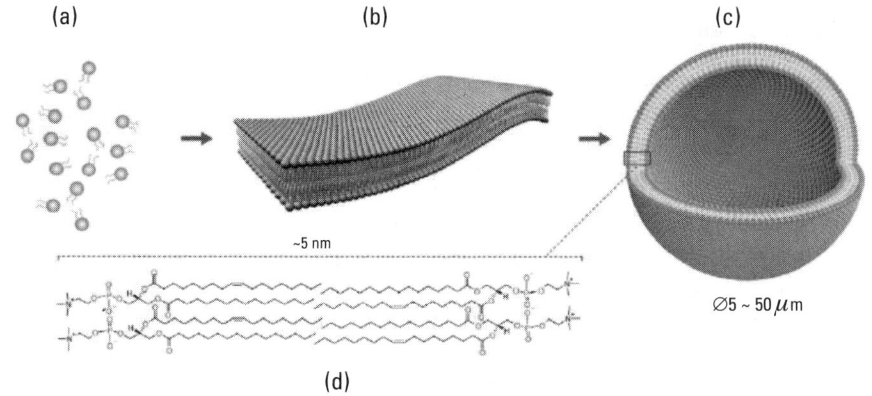

Figure 3.2 Schematic picture of self-assembly of lipid molecules into bilayer membranes and spherical vesicular structures (liposomes). (a) individual lipid molecules in suspension; (b) section of a bilayer membrane; (c) spherical bilayer vesicle; (d) structure of the arrangement of the phospholipid palmityl oleyl phosphatidyl choline (POPC), a main biomembrane constituent.

Particularly interesting as chemical model reactors are giant bilayer vesicles with a diameter >10 μm, since they can be directly manipulated under a light microscope. A number of methods to create giant vesicles have been reported. One of the methods is based on a dehydration-rehydration procedure (Criado and Keller, 1987). As a variant, the rehydration of thin dried films obtained after the evaporation of the solvent from lipid-protein complexes, solubilized in organic solvents, has been used to produce giant proteoliposomes (Darszon, Vandenberg, et al., 1980). The disadvantage of getting a large amount of multilamellar (onion-shell-like) vesicles in the rehydration procedure has been overcome with the electroformation technique (Angelova and Dimitrov, 1986) in which a lipid film is rehydrated in the presence of an alternating current (AC) electric field. Giant unilamellar vesicles (GUVs) with a more homogeneous size distribution around 25 μm in diameter are thus generated. Further improved variants of this procedure yielding larger and more abundant vesicles have been reported, utilizing a spin coating technique for lipid deposition on surfaces (Estes and Mayer, 2005).

Giant liposomes have become the foundation of flexible, soft-walled chemical reactor networks, consisting of a combination of giant unilamellar vesicles containers (GUVs, diameter: 5 ~ 50 μm, internal volume in the pL range), interconnected by membrane nanotubes with a diameter of ~ 100 nm and a broad length range of up to several hundred μm.

3.2.3 Lipid Nanotubes

Besides the liposomes as membrane-enclosed microcompartments, lipid nanotubes, having a diameter in the 10 to 300 nm range, are the other important constituent of nanotube vesicle networks (NVNs). The fundamentals of nanotube formation and the factors which define their physical dimensions are briefly described in the following section. Lipid nanotubes form when a point force is applied to a bilayer vesicle (Heinrich, Bozic, et al., 1999), involving first-order shape transitions, where the thus obtained shape is characterized by a minimum in surface free energy (i.e., the surface-to-volume ratio of the structure is optimized). For a small membrane patch, which is pulled away from a planar surface, the minimum surface area is reached when the point of pulling is connected to the surface by an infinitesimally narrow tether. During the process of pulling, the curvature of the membrane increases as the extending membrane part shrinks towards this tether. Eventually a narrow tube is formed with a radius that is determined by the balance between surface tension and bending rigidity, which prevents the collapse of the extended membrane part to an infinitesimally narrow size. The surface tension σ can be used to describe the interface between a vesicle and its surrounding environment. σ can be employed because only a very small part of the membrane is involved in tube formation, and therefore the term σA contributes to the free energy of the system. Assuming the absence of spontaneous curvature c_0, the curvature free energy for a (perfectly cylindrical) nanotube is written as:

$$F = \sigma A + \frac{\kappa_c}{2}\int (c_1 + c_2)^2 \, dA = 2\pi L_{tube}\left(\frac{\kappa_c}{2R_{tube}} + \sigma R_{tube}\right) \quad (3.1)$$

where σA is the stretching energy contribution and $\frac{\kappa_c}{2}\int (c_1 + c_2)^2$ the bending energy contribution, with membrane surface area A, surface tension σ, bending modulus κ_c, and principal monolayer curvatures c_1 and c_2. L_{tube} and R_{tube} are the length and radius of the nanotube, respectively. The free-energy contribution of the differential stretching is much smaller than the other terms and can be omitted together with the Gaussian curvature contribution, and the contribution of bilayer surface adhesion. The force required for pulling out a tube is:

$$f = 2\pi\sqrt{2\sigma\kappa_c} \quad (3.2)$$

which is typically on the order of a few pN. Importantly, the (steady state) radius of a lipid nanotube, R_{tube}, is a force balance between lateral membrane tension and curvature energy (Evans and Yeung, 1994) and results from minimizing equation 3.1 with respect to *R*:

$$R_{tube} = \sqrt{\frac{\kappa_c}{2\sigma}} \tag{3.3}$$

Because the tube-vesicle system consists of one single continuous membrane, the tube radius can be modulated dynamically in an existing structure by controlling the lateral membrane tension in the vesicle. The energy cost for creating a tube is directly proportional to its length, and a vesicle with a connected tube exists, compared to an untethered vesicle, in a highly tense state. A nanotubular extension can therefore, due to the high curvature, only exist with an adjacent suspension point. A free end would always retract rapidly to the vesicle. This fundamental process is of importance in processes like nanotube-mediated vesicle fusion, which, as discussed in a later section, finds application in multicompartmentalization and the initiation of chemical reactions in vesicles.

Furthermore, due to their very large length-to-diameter ratio in comparison to individual liposomes, nanotubes interconnecting vesicles are most influential to the rate of transport of material between individual containers and, consequentially, on the rates of diffusion-controlled chemical reactions in the network.

3.2.4 Nanotube-Vesicle Networks, Forced Shape Transitions, and Structural Self-Organization

While single unilamellar vesicles are interesting objects for the investigaton of membrane properties, mechanics, and transmembrane transport phenomena (Bagatolli, 2006) (Svenson, 2004), networks of such vesicles constructed by interconnection of individual containers with lipid nanotubes have become an intensively studied area of the biomimetic materials sciences, with particularly intriguing applications as soft-matter micro/nanoreactor systems.

The fabrication of such networks is achieved through a sequence of different transformation processes: As discussed above, self-assembly of individual lipid molecules in aqueous buffered suspension leads to bilayer membrane and vesicle formation (Figure 3.3). Micromanipulator-assisted shape transformations of these vesicles result in lipid nanotube formation, and subsequently additional containers with different internalized chemical

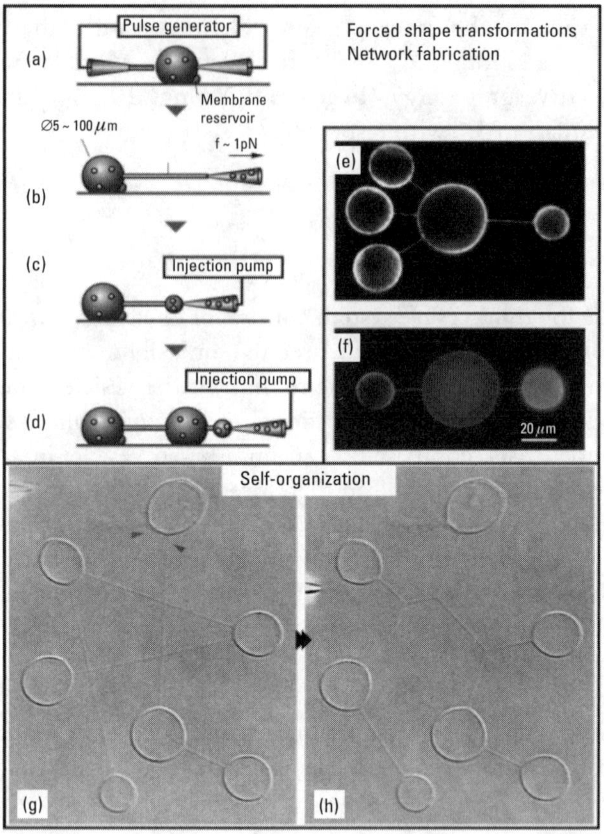

Figure 3.3 Schematic representation of micropipette-assisted formation and internal functionalization of networks of vesicles. A micropipet is inserted into a unilamellar vesicle by means of electroinjection (a), using a carbon microfiber as a counter electrode. The pipet is then pulled away forming a lipid nanotube (b). A new vesicle is formed by injecting the buffer into the nanotube orifice at the tip of the pipet (c). Repeating this procedure can be used to form networks of vesicles. By exchanging the solution in the pipet during the network formation, the interior contents of the vesicles can be differentiated (d). (e) Micrograph of a five-vesicle network with a central container and four daughter-containers connected by single nanotubes. The membrane is stained with a fluorescent dye to enhance visibility. (f) Fluorescence micrograph of a three-vesicle network differentially filled with fluorescent solutions by sequential application of injection pipets filled with different aqueous media. The membrane is also stained with a fluorescent dye. (g) Complex network of seven vesicles, constructed with five unconnected nanotube intersections. (h) The same network of vesicles after stimulated coalescence of nanotubes, leading to a surface free energy optimizing structural rearrangement of the interconnecting nanotubes. The multilamellar reservoir was removed from the structures prior to taking the images. Images reprinted with permission of the American Chemical Society.

environments can be fabricated. With more complex networks, self-organization and geometric self-optimization can then be achieved through mechanical stimulation, leading to further minimization of free-membrane surface energy.

We have developed a toolbox of experimental techniques to control geometry, dimensionality, topology, and functionality in surfactant membrane assemblies for direct application in nanoscale network and device fabrication (Karlsson, Karlsson, et al., 2001) (Karlsson, Sott, et al., 2002) (Sott, Karlsson, et al., 2003). These procedures represent nonconventional fabrication routes to three-dimensional soft-matter devices of differentiated function, featuring geometries at a length scale that is difficult to achieve and a flexibility that is impossible to achieve with modern (solid-state) clean room technology.

Nanotube-conjugated vesicle networks terminated at the ends by surface-immobilized vesicles are most advantageously formed using a microelectroinjection technique (Karlsson, Sott, et al., 2001). This method produces unilamellar (or single-walled) vesicles (10^{-15} to 10^{-12} liters internal volume) interconnected by lipid nanotubes (~25 to 300 nm in diameter) [Figure 3.3(a–d)]. The techniques are related to the variety of experimental micromanipulation techniques developed for membrane tether formation (Derenyi, Julicher, et al., 2002), such as pulling with pipets (Evans and Yeung, 1994), optical tweezers (Dai and Sheetz, 1999), pulling by hydrodynamic force (Borghi, Rossier, et al., 2003), pulling by action of polymers (Fygenson, Marko, et al., 1997), and pulling by action of molecular motors (Koster, VanDuijn, et al., 2003).

To create liposome-lipid nanotube networks, a unilamellar liposome with an internalized volume of a given composition is penetrated by a small

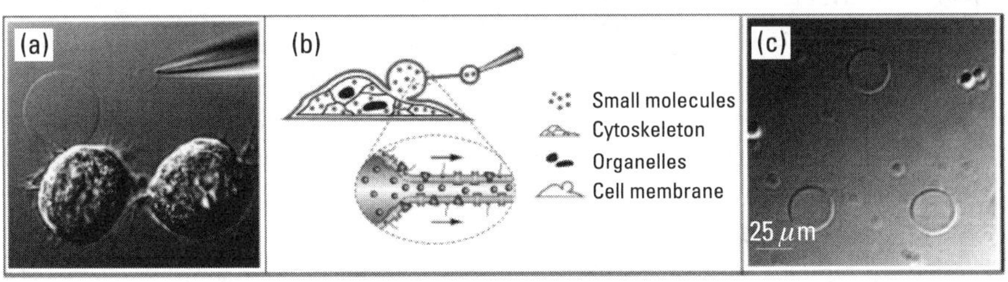

Figure 3.4 (a) Simple two-container network constructed from a cultured NG 108.-15 cell; (b) Schematic drawing of the principle of cell-NVN hybrid structures; and (c) Three-container-NVN constructed entirely from NG 108-15-derived lipid bilayer material. Images reprinted with permission of the American Chemical Society.

buffer-filled glass pipet [Figure 3.3(a)]. Upon retraction of the injection tip away from the liposome, the lipid forms a tight seal around the injection tip. This results in a lipid tube that is connected between the injection tip and the original liposome [Figure 3.3(b)]. By slowly injecting the buffer of a given composition at a rate of tens of femtoliters per second, the nanotube expands at the injection tip, thereby forming a new liposome connected to the original liposome via a lipid nanotube [Figure 3.3(c)]. The composition of the internal volume of the newly formed vesicle can be different from the first vesicle. The extra lipid material required for growth of the liposome is drawn from an attached multilamellar liposome [see Figure 3.3(a)], which acts as a lipid reservoir. In the process, lipid flows from this reservoir along the nanotube to the newly created unilamellar liposome. The flow is required to alleviate the stress imposed on the bilayer membrane when the membrane is expanded at the injection tip and membrane tension is increased. It has been demonstrated that a gradient or difference in membrane tension across a lipid membrane surface drives the lipids to flow from regions of low tension to regions of high tension, in order to eliminate the tension difference (Karlsson, Karlsson, et al., 2002). When the expanding liposome growing on the injection tip has reached a desired size, it can be adhered to a surface. The pipet is then removed and, if required, exchanged to inject a solution of different composition. This procedure is repeated until a network of the desired size, and complexity is obtained [Figure 3.3(d)]. Nanotube-vesicle networks of different connectivity [Figure 3.3(e–h)], internal composition [Figure 3.3(f)] and diverse geometry/topology [Figure 3.3(g, h)] can be created in that manner. A micropipet-assisted electrofusion protocol for formation of networks having complex geometries and higher-order topologies is available (Karlsson, Sott, et al., 2002). Such structures include circularly connected networks as well as networks with three-dimensionally arranged (crossing, but not interfacing) nanotube layers.

A characteristic feature of a fluid-state membrane is the property of structural self-organization and surface-free energy minimization. In the case of nanotube-vesicle networks, surface energy is inhomogenously distributed due to the extreme difference in curvature (by a factor 10 to 100) between the vesicle containers and the lipid nanotubes. In practice this means that nanotubes originating from the same vesicle have to be separated from each other by a distance at the vesicle-nanotube interface in order to preserve the geometry of the system. In Figure 3.3(h), red arrows point to such a constellation (Karlsson, et al., 2002). Otherwise, the tubes will coalesce in order to reduce the surface energy of the system, and form a three-way junction that self-organizes by means of the minimum pathway solution of the specific geometry set by the vertex coordinates of the connected containers, eventually leading to a situation when the angle between tubes equals 120°. As displayed in Figure 3.3(g, h), this self-organization capability can be directly triggered to create structured networks with bifurcating lipid nanotubes. It is initiated by mechanical or electromechanical action, using micropipets or microelectrodes, to merge two nanotubes and is driven by spontaneous minimization of surface free energy. Thus, it is possible to create well-defined complex nanometer-scale subdomain networks with a well-defined geometry and topology within a nanotube-vesicle network. (Lobovkina, Dommersnes, et al., 2004) [Figure 3.3(g, h]).

Besides modification of network structure and internal contents, an equally important feature of the network is their surface integration and attachment. The practical importance of thin, optically transparent substrates for real-time observation and manipulation in inverted microscopy environments results in a somewhat limited choice of available standard materials. To overcome this situation, chemical or physical patterning or surface-treatment techniques can be advantageously employed to common borosilicate glass substrates in order to create substrates with differential adhesion characteristics towards lipid material or vesicles surfaces decorated with biofunctionalized beads (Abdelghani-Jacquin, Abdelghani, et al., 2002). Surfaces such as streptavidin-coated gold films have been used to create defined patterns for vesicle adhesion in which the coordinates for individual nodes can be set with micrometer precision (Sott, Karlsson, et al., 2003). Additionally, techniques for formation of three-dimensional topographic substrates on the basis of high aspect ratio photoresists such as SU-8 have been developed (Hurtig, Karlsson, et al., 2004). Recent reviews provide a detailed introduction into the theoretical and practical aspects of fabrication, interior content differentiation, topology, membrane modification, and surface immobilization of NVNs (Karlsson, Karlsson, et al., 2006) (Karlsson, Davidson, et al., 2004).

3.2.5 Membrane Biofunctionalization of Liposomes and Vesicle-Cell Hybrids

The implementation of complex chemical functions into NVNs requires biochemical functionalization of accessible network components. This can be based on functional materials embedded in the membrane, such as membrane proteins or synthetic multifunctional pores (SMPs) (Sorde, Das, et al., 2003), or the modification of the vesicle's interior volume, for example by compartmentalization or internalization of functional (bio)-polymers.

For example, to enable communication between the interior and exterior of a vesicle-nanotube network, providing exact means of external control over processes inside a container network, mediation of a chemical signal through the membrane is of foremost importance. Using membrane proteins (i.e., transporters), and channels as highly selective signal mediators due to inherent molecular recognition and permeability properties, selectivity over the transported chemical species can be achieved. Many membrane proteins also have an additional gating function controlled by membrane potential or specific ligands. Membrane-based devices of this kind can be readily used where a series of manipulations have to be performed on an initially extremely small volume containing a signalling molecule or an analyte in analytical, biosensor, and computational devices. By combining biological and micromanipulation techniques, a new method for producing NVNs using reconstituted proteins has also been developed. Membrane-spanning proteins such as ion transporters can be introduced to the networks either in their surrounding membrane or as integrated parts of organelles or catalytically active particles, confined in individual containers (Davidson, Karlsson, et al., 2003).

As a distinct advantage over all other ultra-small-scale reactor concepts with respect to integration with biological components, NVNs already possess a biomimetic bilayer membrane as the main system boundary. This offers, in principle, the opportunity to obtain membrane proteins and lipids from natural sources in high yields, while fully maintaining their functionality. The function of membrane constituents and specifically bioconjugated natural lipids can therefore be transferred directly to the NVNs, either as a practical way to study a sensitive biomaterial (e.g., a membrane protein in a simplified environment), or to create biomimetic devices with a specifically selected function.

Generally, a wide range of methods have been developed for reconstitution of membrane proteins in artificial SUVs (Eytan, 1982) (Rigaud, Pitard, et al., 1995) and subsequently GUVs, that can be used to build networks (Davidson, Karlsson, et al., 2003). Reconstitution in GUVs include fusion of proteoliposomes generated by detergent-mediated reconstitution or insertion of proteins into preformed GUVs via peptide-induced fusion (Doeven, Folgering, et al., 2005) (Girard, Pecreaux, et al., 2004) (Kahya, Pecheur, et al., 2001). This approach has certain limitations. First, the proteins have to be removed from their native environment, usually by application of detergents, where it often is not possible to ensure that all the proteins are reconstituted uniformly in their desired orientation. Secondly, as the isolation procedures can be quite harsh, protein activity may be reduced or completely lost. Experimental work in this field can be time-consuming and difficult to reproduce, as suitable reconstitution protocols for each desired protein need to be developed. The most beneficial approach with respect to these problems is the formation of the NVNs directly from a native cell membrane, which is unfortunately a complicated endeavor as it is firmly attached to the cytoskeleton.

Simplified access to the cell membrane can be obtained by exploiting natural reactions of biological cells to particular means of stress. Upon chemical or mechanical stress (Cunningham, 1995) (Harris, 1990) (Schutz and Keller, 1998) (Rentsch and Keller, 2000) (Zhang, Gao, et al., 2000), cytokinesis (Fishkind, Cao, et al., 1991) (Burton and Taylor, 1997), cell movement (Friedl and Wolf, 2003) (Trinkaus, 1973), as well as during apoptosis (Mills, Stone, et al., 1998), cells can form unilamellar micron-sized protrusions, also known as membrane blebs. These structures are compatible with the micromanipulation procedures and tools used for synthetic vesicles (Karlsson, Sott, et al., 2001) and serve as readily available precursors for NVN formation. The cell presumably still maintains its full functionality while adapting to external stress factors by continuously modifying cytoskeletal adhesion to the plasma membrane. If there is a local defect in membrane-cytoskeleton attachment, a bleb is extruded by inflation of the detached membrane by intracellular fluid flow. The total cell volume, however, stays nearly constant.

In order to utilize membrane blebs for hybrid structures combining the cellular plasma membrane with vesicle networks, bleb formation can be triggered in various ways. The preferred chemical method is based on using a combination of dithiothreitol (DTT) and formaldehyde. Membrane blebs can be used to form surface-adhered networks of plasma membrane vesicles, with typical vesicle diameters of 5 to 10 μm and tube lengths of several tens of micrometers, using the electroinjection technique developed for NVN fabrication. This method provides high yields, native composition, correct protein orientation, and function. Only about 250 μm^2 of membrane area is needed to build a three-vesicle network (given a 5-μm vesicle diameter and 40-μm total tube length), thus such a network can be practically built from a single cell. Whereas it appears likely that most cell-membrane components can be extracted into a bleb, the subsequent morphological shape change and the preceding cellular processes may cause a different composition of the bleb membrane compared to the native membrane (Keller, Rentsch, et al., 2002). For instance, membrane proteins that are interacting with the cytoskeleton may not move over to the protruding bleb, and a change in lateral diffusion of proteins may cause an altered spatial membrane protein distribution (Tank, Wu, et al., 1982).

Figure 3.4(a) displays a micrograph image of an adherent cell with a protruding bleb, connected via a nanotube to a daughter vesicle which is still attached to the micropipette. Following the same general protocol developed for NVN construction, the cell-derived bleb is used as a source for formation of a daughter vesicle. In Figure 3.4(b), a schematic representation of the method is shown. Organelles and cytoskeletal structures remain in the adherent cell, while the bleb most likely encloses low-molecular-weight cytosolic components. The bleb and the subsequently formed daughter vesicle have membrane proteins embedded in the membrane that are properly oriented and fully functional. However, larger structures such as organelles, if present in the mother bleb, can not pass through the nanotubes because of size constraints. Using this method, networks of different connectivity entirely derived from bleb membrane material can be constructed. Figure 3.4(c) shows a brightfield micrograph image of a network of three vesicles, linked to each other by a three-way nanotube junction. The presence of membrane glycoproteins in the network boundary was verified in this particular case using a selectively binding dye (WGA-Alexa, 488), with affinity to their sialic acid and N-acetylglucosaminyl residues (Bauer, Davidson, et al., 2006).

3.2.6 Internal Volume Functionalization and Compartmentalization of Nanotube-Vesicle Networks

A relatively young field of research is the dynamic compartmentalization of liposomes. Only very few examples of successful introduction of size- or shape-controllable subcompartments into vesicles have appeared. (Bolinger, Stamou, et al., 2004) (Long, Jones, et al. 2005) (Markstrom, Gunnarsson, et al., 2007). The development in this area is driven mainly by the desire to increase complexity and versatility and to reach more realistic artificial cell models. The complexity of a biological cell is very much related to the existence of dynamic internal functional compartments, a feature that is normally absent in vesicular systems. Therefore multiple and, ideally, reversible compartmentalization in artificially created micro- and nanocontainers, such as giant unilamellar vesicles, is of growing interest for the construction of artificial cells, both from industrial and basic research perspectives (Pohorille and Deamer, 2002). Particularly, a combination of materials that can be selectively incorporated and liberated from the compartments leads to novel model systems with high potential for drug release, the initiation of chemical reactions and separation/fixation of biologically relevant macromolecules

(Jesorka, Markstrom, et al., 2005). A strong motivation for the investigation performed in our group is to eventually utilize dynamically compartmentalized GUVs as structurally complex, hierarchically organized chemical reactors.

Compartmentalization is used by the living cell to isolate chemically and physically dissimilar environments from each other, allowing different, incompatible metabolic activities to occur simultaneously. An experimental model for cytoplasmic organization using a PEG/dextran aqueous two-phase system (ATPS) to create heterogeneous protein distributions within liposomes has been reported (Long, Jones, et al., 2005). Unfortunately, this method is currently limited to the creation of one single compartment of random size [Figure 3.5(a)]. A high degree of flexibility is possible through the utilization of "smart" materials, such as thermoresponsive polymers with the ability to generate dense hydrogels at higher temperatures. Temperature-induced, fully reversible and therefore easily controllable compartmentalization is achieved using the hydrogel-forming lower-critical-solution-temperature (LCST) polymer poly(N-isopropyl acrylamide) (PNIPAam) in the internal volume of a NVN. In a high ionic strength buffer, an in-vesicle formed hydrogel collapses into a hydrophobic compartment that reaches a density which is typical for macromolecular crowding in living cells (Ellis, 2001-1). Such compartmentalization and density control in NVNs is an

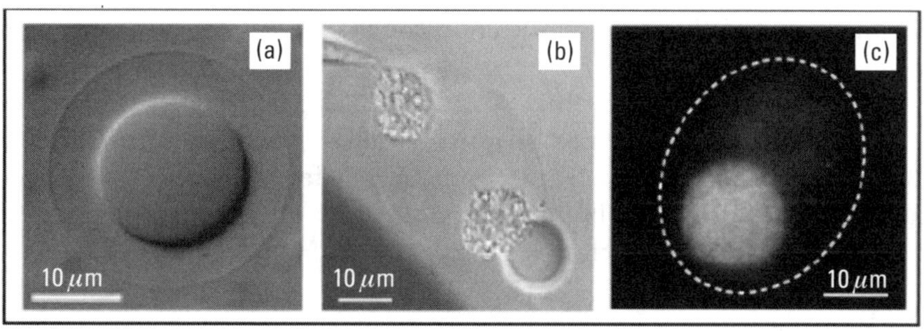

Figure 3.5 Dynamic compartmentalization of giant unilamellar vesicles as a means of internal volume functionalization. (a) Dynamic (PEG)/dextran aqueous two-phase system with a single internal compartment obtained by phase segregation inside a GUV. (Long, et al., 2005, reprinted with Permission of PNAS.) (b) Multiple dynamic microgel compartments within a single vesicle obtained from the thermoresponsive poly(N-isopropylacrylamide, PNIPAam) upon heating above LCST.) (Image reprinted with permission of the Royal Chemical Society.) (c) A vesicle with two PNIPAam compartments above LCST, each with different fluorescent particles embedded within the compartment.

important prerequisite for the modeling and study of truly cell-like reaction rates and mechanisms.

The multicompartmentalized vesicles in Figure 3.5(b, c) are obtained by a nanotube mediated vesicle fusion procedure. A nanotube-interconnected two-vesicle network is injected with PNIPAam in aqueous buffer, and upon heating the system, a hydrogel is formed inside each vesicle. This can rapidly and conveniently be done on-chip with surface-integrated resistive heating structures, consisting in this case of a thin evaporated gold layer. The dark lower-left corner in the micrograph in Figure 3.5(b) is the (less transparent) area covered by such a surface-printed heater. The hydrogel collapses eventually, and any coinjected material that is either larger than the pore size of the collapsed gel or has an affinity to the gel-forming polymer (e.g., nanosized beads or DNA), is trapped inside. Subsequently, both vesicles are joined through tension-driven nanotube retraction, leading to hydrophobic subcompartments of different composition in a single vesicle. Figure 3.5(b, c) shows containers with two differentiated internal gel-compartments created by this fusion procedure. In the vesicle depicted in Figure 3.5(c), two kinds of fluorescent nanoparticles are entrapped in individual gel compartments.

The method is easily expandable to more than two compartments, and several directions are, in principle, open to increasing the complexity of the system even further. Surface-functionalized beads, nanometer-sized small unilamellar vesicles, or functional side groups on the polymer chains allow for a wide range of chemical functionalities to be introduced. The LCST polymer system is fully reversible; upon cooling below LCST the compartments dissolve, and release their entrapped contents. Permanent fixation of compartments could be realized by cross-linking a specifically designed LCST polymer in the collapsed gel state. Different gel compartments are created independently from each other and can therefore contain functionalized material that is chemically interreactive, even after compartments are combined in the same vesicle.

3.3 Transport Phenomena in Nanotube-Vesicle Networks

In nanotube-vesicle networks, locally internalized (injected or generated) materials can travel through the network structure. This mobility is governed by different, passive diffusional or active, tension or field-induced means of transport. Due to their small size and structural flexibility, NVNs allow for controlled transport of ultrasmall amounts of material down to the single-molecule level. Size and optical properties of the networks (essentially transparent in the visible and near UV region) mostly allow for direct monitoring of transport processes, provided fluorescent or fluorescently tagged molecules are employed, as well as determination of the mobility, properties, and chemical reactivity of transported species. The most important mode, also from a biological perspective, is passive transport by diffusion. Particularly, the understanding and characterization of mixing and diffusion processes within such confined spaces, as laid out in this section, have become the foundation of our investigations of biochemical reactions, as will be discussed in the final section.

3.3.1 Mass Transport and Mixing in Nanotube-Vesicle Networks

In order to utilize vesicle-nanotube networks as small-scale chemical reactors (e.g., for enzymatic reactions or for studies of reactions in a crowded cell-like environment), well-defined transport of materials and solutes through the nanotubes connecting the vesicle containers is essential. Three fundamental mechanisms of material transport within lipid nanotubes have been investigated to date. The first, Marangoni transport, is based on membrane tension gradients and utilizes the dynamic and fluid character of the bilayer membrane. The second mechanism is electrophoresis, a well-established way of moving fluids and solutes in micro- and nanofluidic devices. The third, and practically most important mode, is based on diffusion, a very effective means of transport over short distances. Figure 3.6 gives an overview of the transport modes, always assuming that the size of the moving species is considerably smaller than the nanotube size.

Due to the nature of the fluid membrane, different velocities have to be taken into consideration: the velocity of the solvent in the tube (V_S), the velocity of the transported molecule or nanoparticle (V_P), and the velocity of the membrane itself (V_M). In Figure 3.6, for each transport mode the relationship of these velocities to each other is denoted.

Brightfield micrographs of simple networks utilized in the investigation of the respective transport modes are shown in Figure 3.7. The networks in each case were constructed using the general procedure described in Section 3.2, with the multilamellar vesicles removed. In all schematic drawings, the size relation between tube and container is not realistic. True size relations are visible in the images in Figure 3.7.

In Figure 3.7(a) a two-container network for diffusive transport is shown, in Figure 3.7(b, c) a two-vesicle network with a small particle entrapped in the interconnecting tube, utilized to demonstrate the membrane coupled transport upon mechanical creation of a tension gradient. The

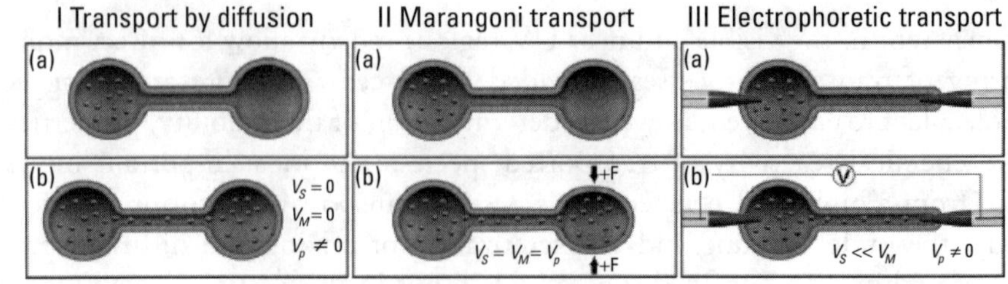

Figure 3.6 Three different mechanisms for transport through a nanotube. I: Transport by diffusion. I(a): Schematic drawing of a cross-section of two vesicles connected by a lipid nanotube. I(b): Schematic representation of diffusional transport, driven by the concentration gradient between the two vesicles. II: Tension-driven (Marangoni) lipid flow. II(a): Schematic of a cross-section of two vesicles connected by a lipid nanotube. The system is at rest. II(b): Schematic showing Marangoni transport. The right vesicle is deformed, creating a tension gradient in the network. Fluid and particles trapped in the nanotube are transported in the direction of the gradient. III: Electrophoretic transport. III(a): Schematic of a cross-section of a vesicle connected to a pipet via a lipid nanotube. III(b): In an electric field across the lipid nanotube, charged species, such as DNA, are driven through the nanotube. (Images I-III(a) represent a snapshot immediately after release of particles in the left vesicle; V_S = solvent velocity, V_M = membrane velocity, V_P = velocity of a particle being transported.)

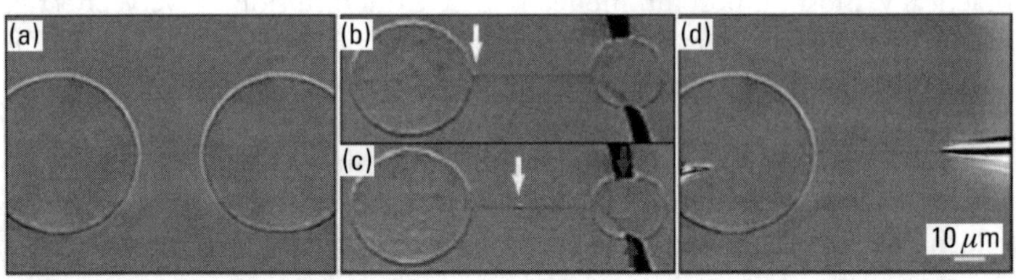

Figure 3.7 Differential interference contrast (DIC) microscope images of experimental arrangements for the investigation of the three transport modes, (a) Diffusion transport, (b, c) Marangoni-transport, (d) Electrophoretic transport. Gray arrows indicate the mechanical force exerted on the membrane to create tension, which leads to displacement of the particle entrapped in the nanotube. (The shift in position is indicated by white arrows.) Images reprinted with permission of the American Chemical Society.

particle, indicated by white arrows, travels along the tube in the direction of the gradient, as membrane is moved to reduce tension at its source, indicated by gray arrows in the figure. Figure 3.7(d) shows the basic setup of an electrophoresis experiment, with two electrode-pipets contacting on one side a vesicle and on the other side the opening of a suspended lipid nanotube. The electrodes are outside the visualized area.

3.3.2 Transport by Diffusion

In macroscopic systems, diffusion is generally slow and only of limited use for moving solutes and material. When approaching the micro- and nanoscale region, diffusion becomes an efficient transport mechanism for moderately sized molecules or particles. The average time for a molecule or a colloidal particle to diffuse over a distance L is given by:

$$\tau = \frac{L^2}{D} \tag{3.4}$$

D being the diffusion coefficient. D is inversely proportional to the size of the molecule or particle, and therefore large entities such as polymers or proteins diffuse significantly slower than small molecules. Equation (3.4) also sets the timescale for equilibrating solute concentration gradients across a container of size L, for example, when mixing two solutes without agitation. If it is desired that the vesicle contents remain differentiated, diffusion through the nanotube is unwanted, and care must be taken in the design of the network and in the choice of solutes. However, diffusion might also be used as a means of active transport between vesicles in the networks.

NVNs have three unique properties that make diffusion interesting as a means of transport. They are sufficiently small; hence, diffusion is effective as a mode of transport. Furthermore, the chemical potential in the networks can be controlled over time by injection of different concentrations of relevant species into individual containers. Finally, the geometry of the networks can be changed over time. Thus, the directionality of transport can be precisely controlled. Reactants can therefore be transported from one end to the other; and sequentially catalyze reactions in different nodes as has been shown for diffusive directed transport of alkaline phosphatase (Sott, Lobovkina, et al., 2006).

For diffusive transport in nanotube-vesicle networks, it is important to mention that the diffusion time given by (3.4) is not necessarily the most relevant. It merely gives the timescale for establishing a concentration gradient over a tube of length L. Once this gradient is established, it takes a much longer time to transport the content of one vesicle to another, due to the small dimension of the nanotube. Diffusional transport in a nanotube can be viewed as a one-dimensional motion. The equilibration time for concentration differences between nanotube-conjugated vesicles has been calculated in (Dagdug, Berezhkovskii, et al., 2003), in its simplest form the relaxation time is

$$\tau_{relax} = \frac{VL_{tube}}{2\pi D r_{tube}^2} \tag{3.5}$$

where V is the volume of the nanotube-conjugated vesicles. The ratio between this relaxation time and the diffusion time τ for a particle through the tube as given by (3.6), takes a simple form.

$$\frac{\tau_{relax}}{\tau} = \frac{V}{2V_{tube}} \tag{3.6}$$

The relaxation time is determined solely by the ratio of the vesicle to tube volume. Clearly this ratio is always large, and it can be controlled by adjusting the size of the vesicles (pipet injection) and dimensions of nanotubes (choice of membrane material and tension), thus allowing to fully control whether diffusive transport is to be used or avoided.

A specifically interesting case is the diffusional transport in larger NVNs. Recently, a generic model based on particle diffusion has been developed (Lizana and Konkoli, 2005), making a step forward towards understanding the material transport properties of complex networks, where geometrical concepts such as the lengths of the nanotubes play an important role. In complex networks composed of a larger number of containers of different sizes and of lipid nanotubes of different lengths and connectivity, the time-dependent concentration variation of molecular species in each separate container is not trivial to predict. However, one can derive a set of equations that describe how the number of particles in each vesicle evolves over time. The problem is approached numerically based on the geometry of the network, its connectivity, and the bulk diffusion coefficients of the molecular species. By varying the initial conditions it is possible to induce wave-like concentration behavior, with concentration maxima and minima, in selected vesicles. This approach has been exploited in reaction-diffusion networks, in which it is possible to predict the time-dependent concentration changes of several species; enzyme(s), substrate(s), and product(s), in each container in the network. For example, the transport in the networks exhibits a distinct sensitivity to geometrical changes in the structure. The model is expandable in many ways (e.g., to more sophisticated intracontainer dynamics), or allowing particles in the system to be disturbed by each other.

Experiments have been performed in which an enzyme is allowed to diffuse through linear and branched networks filled with substrate. As an example, the diffusion of fluorescein (100 μM) through a network is displayed in Figure 3.8. The figure shows a three-vesicle network homogenously filled with fluorescein, where two of the vesicles (2 and 3) were photobleached by laser illumination (488 nm). The diffusion of fluorescein from vesicle 1 into the rest of the network was then monitored over time [Figures 3.8(b, c)]. The fluorescence intensity in the nonbleached vesicle decayed rapidly, and the fluorescence signal in all vesicles eventually converged to the same value, indicating an even distribution of fluorescein throughout the network [Figure 3.8(d)]. The poor recovery of the fluorescence in vesicle 2 and 3 is caused by both photobleaching and leakage of fluorescein.

The experimental results coincide very well with the theoretical description (Karlsson, Sott, et al., 2005). The scope of these experiments is extended further in Section 3.4, where chemical reactions in NVNs and control of enzyme diffusion dynamics by network geometry and initial conditions are described in detail.

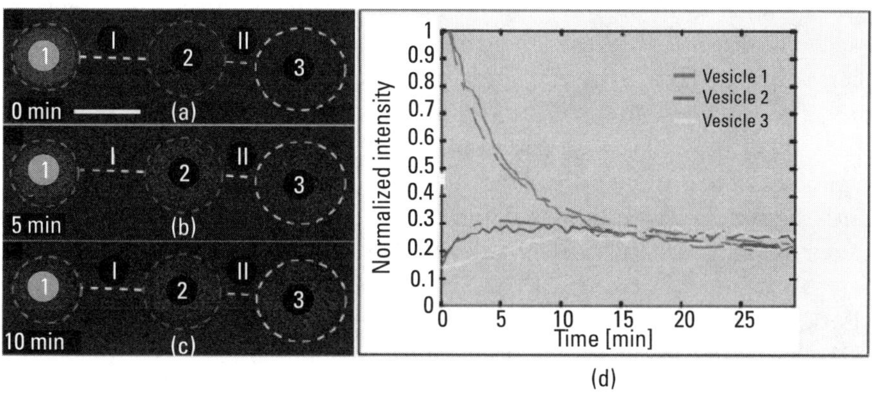

Figure 3.8 Diffusion of fluorescein through a NVN. (a–c) Confocal LIF-images of an NVN containing fluorescein. The boundary of the vesicles and the connecting nanotubes are marked with dashed lines. (d) Graph showing normalized fluorescence intensity plotted versus time. The colours of each curve correspond to the color of the dashed line around the vesicles in panels (a–c). Dash dotted lines show theoretical values of the product diffusion. Fluorescence images were digitally edited to improve image quality. The scale bar in panel (a) represents 10 μm. Images reprinted with permission of the American Chemical Society.

3.3.3 Tension-Controlled (Marangoni) Lipid Flow and Intratubular Liquid Flow in Nanotubes

The bilayer membrane is in a fluid state (liquid crystal), and behaves as a two-dimensional liquid. In a system of two fluids separated by an interface, the interfacial tension must be uniform when at equilibrium. However, in nonequilibrium situations, a tension gradient can exist along the surface of a fluid system. In general, this results in a convective flow. Interfacial tension-driven flows are generally called Marangoni flows. Common examples of this flow behavior are the spreading of liquid films on a surface due to temperature gradients, Bernard-Marangoni convection cells, and flows due to surfactant concentration gradients (Nepomnyashchy, Velarde, et al., 2002).

In surfactant bilayer membranes, tension gradients can be produced by various means, the most direct way is through mechanical perturbation of the membrane. Other means of creating surface tension gradients include bulk hydrodynamic flows, temperature gradients, electric fields, highly focused laser spots, aspirating lipids (Bar-Ziv, Moses, et al., 1998), or solute gradients inducing osmotic pressure gradients.

When the membrane tension is increased at a point in the membrane, the lipid membrane system will respond to this perturbation by transporting lipid material from regions of lower tension to the point of higher tension. Mechanical perturbations using microprobes can be utilized to locally increase the surface-to-volume ratio in vesicles conjugated by nanotubes and drive flow in a desired direction (Karlsson, Karlsson, et al., 2002) (Karlsson, Karlsson, et al., 2003). The establishment of a tension gradient is very rapid, typically on the order of milliseconds, followed by a steady flow of lipids until the system has equilibrated, which typically takes several seconds. If the system is being gradually perturbed, a flow can be sustained for much longer.

A theoretical description for Marangoni flow in a simple network (two vesicles connected by a nanotube) has been developed by Dommersnes et al. (Dommersnes, Orwar, et al., 2005). The flux of liquid driven by Marangoni flow is given by

$$J_M \propto \frac{r_{tube}^3}{\eta} \frac{d\sigma}{dx} \tag{3.7}$$

where r is the nanotube radius and $d\sigma/dx$ is the tension gradient along the nanotube.

Since the tension gradient has been achieved through the deformation of a vesicle, the hydrostatic pressure in that vesicle has increased. The pressure difference between vesicles will thus give rise to a Poiseuille flow given by

$$J_P = \frac{\pi r_{tube}^4}{8\eta L} \Delta P \tag{3.8}$$

where L is the length of the nanotube and ΔP is the pressure difference between two vesicles.

For large vesicles, the internal pressure is extremely low and the Poiseuille flow is dominated by the Marangoni flow. In general the balance between Marangoni and Poiseuille flow can be found from the following approximation

$$\frac{J_M}{J_P} \propto \frac{R_v}{r_{tube}} \approx 10^3 \tag{3.9}$$

where R_v is the vesicle radius. The expression states that as long as the radius of the vesicle is much larger than the nanotube radius, the Marangoni effect will be the dominant contribution to transport in nanotubes.

When a tension difference is applied between two vesicles connected by a nanotube a tension gradient is established very quickly with a relaxation time $\tau_\sigma \sim 1$ ms. This is associated with stretching out the small variations in lipid density. As this happens, the tension gradient increases (Dommersnes, Orwar, et al., 2005). This is a transient state where the radius of the nanotube is unchanged. As the lipid membrane is stretched, the nanotube starts to change shape. This is called the peristaltic mode where fluctuations in tube diameter relax with a relaxation time τ_r until the tube reaches the final stationary state as long as the tension gradient is kept constant. This second relaxation time can be thousands of times longer than τ_σ, meaning that while the tension gradient relaxes very fast, the tube shape changes slowly, on the order of seconds.

Importantly, the velocity of solvent inside the tube is practically equal to the velocity of lipid flow (i.e., there is a Marangoni plug flow of solvent inside the tube), in contrast to the parabolic Poiseuille flow found in solid-state channels. Therefore, in a more complex network transport can be achieved between two selected containers without affecting the rest of the network. A two-point perturbation technique has been developed, where the membrane tension in one surface-immobilized vesicle (from which material is to be taken) is decreased at the same time as the membrane tension in another surface-immobilized vesicle (to which material is to be transported) is increased (Karlsson, Karlsson, et al., 2003). With this two-point perturbation technique, a difference in membrane tension between the two targeted containers can be created, that is much larger than for all other connected containers. Consequently, during the manipulation, material transport is only taking place between these two containers. The technique opens up the possibility to use tension-driven lipid flow in selected regions of complex multicontainer networks.

3.3.4 Electrophoretic Transport

Electric fields can also be used to transport solutes and particles through nanotubes connecting liposomes in a network (Tokarz, Akerman, et al., 2005). Micropipettes with inserted Ag/AgCl electrodes were applied to maintain an electrophoretic potential across the network. In order to suppress electroosmotically driven volume flow through the pipette tips, the surface of the pipettes was modified with hexamethyl disilazane (HMDS), a hydrophobic agent, and the pipette tips were capped insitu with 20% cross-linked acrylamide gel. Since the lipid membrane is negatively charged, it is pulled by the electric field towards the positive electrode. The movement of the lipid membrane will also cause the liquid contained inside the tube to move along it. However, since an electric field is also applied inside the tube, the resulting electroosmotic flow will counter the membrane-driven liquid flow to a large degree. Approximately, there is a 10% difference between these flows, the resulting velocity profile of the fluid being dominated by the electroosmotic flow at the walls and by the Poiseuille flow in the middle of the tube. The electric field is primarily applied inside the tube, and due to the low conductivity of the membrane, the field at the nanotube exterior surface is extremely small.

This technique in combination with confocal microscopy and sensitive fluorescence detection, has been used by Tokarz et al. to transport and detect large (5.4 to 166 kbp) double-stranded DNA molecules with a detection efficiency close to unity. The size dependence of the DNA conformation inside the nanotube was elucidated from the fluorescence bursts originating from the individual DNA molecules as they are detected during transport (Tokarz, Akerman, et al., 2005).

3.3.5 Solution Mixing-in Inflated Vesicles through a Nanotube

In the process of constructing NVNs consisting of vesicles with different content, mixing of internal solution from a vesicle A with newly pipet-injected material into a growing vesicle B, has also to be taken into consideration (Figure 3.9). During injection, a membrane flow originates from the tension gradient that is being created when vesicle B grows, drawing lipid material from an attached multilamellar vesicle (membrane reservoir). This membrane movement creates a coupled flow of solution A from vesicle A through the nanotube into vesicle B.

Thus, if solution B in the injection needle and internalized solution A in vesicle *A* (from which material is extracted through a nanotube) are different, mixing of the two solutions will take place in the growing vesicle B. The growth of the vesicle generates a Marangoni flow of surfactants in the nanotube which in turn generates a flow of solvent A inside the nanotube, counter-directional to the pressure-injected solvent B. The volume ratio ψ between solvent *A* and *B* inside the mixing vesicle depend only on geometrical quantities. In the simplest case it is:

$$\psi = \frac{V_B}{V_A} = \frac{3r}{2R} \tag{3.10}$$

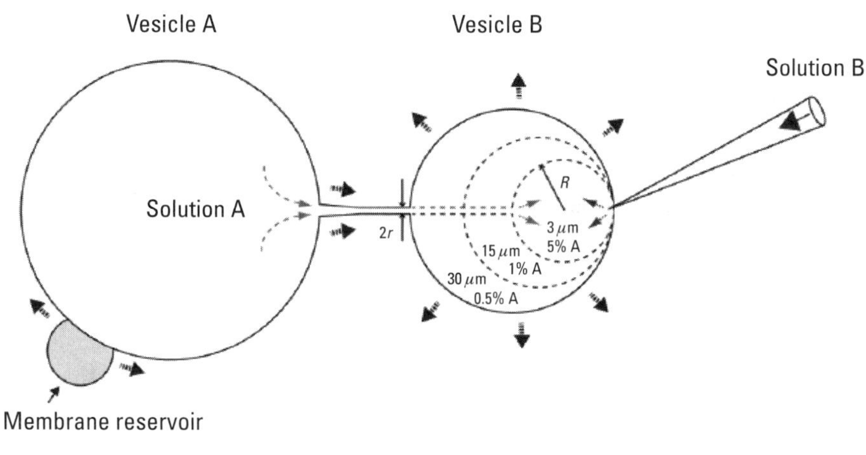

Figure 3.9 Membrane-coupled flow leads to mixing of internalized content of vesicle A with the injected material-filling vesicle B. The mixing ratio depends only on the radii R and r of vesicle b and the interconnecting nanotube, respectively.

where r is the nanotube radius and R is the radius of vesicle B. This equation is valid as long as $R >> r$ under plug flow conditions. The mixing ratio only depends on geometrical quantities. With a tube radius $r = 100$ nm, and vesicle radius $R = 3\ \mu$m, there will be 5 % of solvent A in the mixing vesicle, at $R = 15\ \mu$m 1% of solvent A and at $R = 30\ \mu$m 0.5% of solvent A. Thus, the majority of fluid injected to the growing daughter vesicle comes from the pressure-based injection, and for large vesicle sizes it dominates. In this system, if material dissolved or suspended in solution A is smaller than the nanotube diameter ($2r$), it will be carried along into vesicle B by the solvent flow. If it is larger, such as colloidal particles and organelles, it will be retained in vesicle A.When several vesicles are formed sequentially, controlled mixing can occur, spanning several orders of magnitude (Davidson, Dommersnes, et al., 2005).

An alternative retraction/mixing method has been developed by Karlsson et al. A vesicle which has been inflated from a nanotube but still attached to a pipet can be released by applying an electric pulse through the pipet opening. The network immediately seeks to minimize its surface energy by retracting the nanotube between the immobilized and the released vesicle. The released vesicle is pulled toward the target vesicle until both containers merge. The contents mix by diffusion, which at the short length scales in question is a very fast process. This procedure has been successfully used to initiate enzymatic reactions in lipid vesicles (Karlsson, Sott, et al., 2005), to be described in the following section.

3.4 Chemical Reactions in Nanotube-Vesicle Networks

In chemical reactors with a characteristic length scale of micro- and nanometers, the understanding of reaction kinetics is of increasing interest. Since at this small scale, diffusion is the predominant mode of transport and mixing, the interplay between reactions and diffusional transport is of special interest.

The noncompact nature of structured NVNs makes the situation rather complicated, and calls for a specifically adapted theory. In the first part of this section, an introductory review of the situation is given, while in the remaining parts experimental details of control and initiation of chemical reactions in NVNs are presented.

Central to the discussion about the differences between small scale and large scale reactions is the question about which reactions benefit from structured spaces. There are a large number of reactions that run faster in a network-like geometry. The interplay between geometry and the nature of a reaction scheme becomes important only when the individual reaction steps start to influence each other. Such reactions contain antagonistic catalytic influences in the intermediate stages of a multistep reaction scheme. This approach is, for example, an alternative way to explain certain aspects of cytoarchitecture. In mitochondria, it is assumed that the structured nature of the inner membrane is a way of increasing surface area, but an additional mechanism, the beneficial influences of structured space on antagonistic catalytic influences, is possible (Konkoli, 2005).

Due to their versatility, flexibility and favorable materials properties, vesicle-nanotube-networks present a potent system to investigate chemical reactions in confined and structured space and under macromolecular crowding conditions. Reaction conditions can include interactions with biomacromolecules and cell components in their native environment, while the fluid character of the membrane offers practical advantages, such as facile interfacing to injection equipment and active surfaces.

3.4.1 Diffusion-Controlled Reactions in Confined Spaces

In general, diffusion-controlled reactions (DCR) are ubiquitous in nature. They are central to a broad range of phenomena starting from matter antimatter annihilation in the early universe down to the ultra-small scale reaction compartments in the interior of the living cell. Diffusion is an efficient means of transport when the distance to be traveled by molecules is rather small; therefore the concept of diffusion is frequently employed in the chemical reaction dynamics of biochemical reaction pathways.

In this section an introductory review of general and theoretical aspects of diffusion controlled reactions is given, placing special emphasis on the relevance for chemical kinetics in nano-environments (ultrasmall volumes) that are engineered, for example fluid membrane nanoreactors, or are present in the living cell (Golgi apparatus, mitochondrion, or endoplasmic reticulum).

The behavior of DCRs in large volumes has been studied intensively in the past. They are rather well understood and extensive material is available concerning the development of experimental tools (Kopelman, 1988) (Yen, Koo, et al., 1996) and theoretical methods (Kotomin and Kuzovkov, 1992) (Kotomin and Kuzovkov, 1996) (Ovcinnikov, Timasev, et al., 1989) to describe such systems. The main goal has been to design equipment to measure particle distributions both in space and time and to develop consistent theoretical frameworks to describe and predict their behavior.

To describe the problem, a standard procedure can be used where the reaction volume is divided in small units/cells C_1, C_2, ..., C_x, ..., C_L (Konkoli, 2004). The choice of the number of cells L is governed by accuracy requirements; normally a very large L is taken. A configuration of the system is specified by a set of cell occupation numbers $\mathbf{n} = (n_1, n_2, \ldots, n_x, \ldots, n_L)$ that change in time in a stochastic manner. The goal is to compute the probability p($\mathbf{n}$, t) to find the system in configuration n at time t. Starting from microscopic reaction rules, it is possible to construct a master equation that governs the dynamics of the system:

$$\dot{p}(\mathbf{n},t) = \sum_{\mathbf{m}} W_{m\to n}\, p(\mathbf{m},t) - \sum_{\mathbf{m}} W_{n\to m}\, p(\mathbf{n},t) \tag{3.11}$$

where $\dot{p}$ denotes the time derivative and $W_{m\to n}$ denotes a rate of transition from state m into state n. In practice, due to the large number of involved states, this master equation is difficult to solve, as it scales exponentially with L. Instead of solving (3.11) directly, one can focus on the particle concentration:

$$\rho(x,t) = \sum_{\mathbf{n}} n_x\, p(\mathbf{n},t) \tag{3.12}$$

Even though this is a frequently employed route, new problems emerge: the equation that describes the dynamics of $\rho(x, t)$ involves a higher order density function

$$\rho(x,y,t) = \sum_{\mathbf{n}} n_x n_y\, p(\mathbf{n},t) \tag{3.13}$$

Such behavior continues indefinitely, leading to an, in principle, infinite hierarchy of equations. This set of equations (not presented here since it is model dependent) has to be truncated to be successfully employed. DCRs are an example of an extremely complicated many-body problem, and truncating a set of these equations is far from trivial.

When the system size is infinite, diffusion is a rather inefficient means of transport. Due to the presence of reactions and the slowness of diffusion, large spatio-temporal variations in occupation numbers n may develop in time. Any attempt to describe the problem using a mean-field formalism is bound to fail. Mean field treatment amounts to neglecting information on fluctuations by assuming $\rho(x, y, t) \approx \rho(x, t)\rho(y, t)$. As a rule, this approximation tends to fail below some critical dimension, which is problem dependent and can be as low as one or higher than three, thus indicating that dynamics is always fluctuation dominated. The reaction A + B → C is an example of a case where the critical dimension is four. Kinetics of this type are normally referred to as *fluctuation dominated kinetics.*

Fluctuation effects are important in the context of DCRs. To describe their dynamics correctly, information on higher order density functions has to be obtained. In practice, one keeps track of a few lowest-order correlation effects, that is, focusing on pair-correlation effects described by $\rho(x, y, t)$ while neglecting higher order correlations, such as by taking $p(x,y,t) \approx p(x,t)\,p(y,t)\,p(z,t)$. It is hard to foresee under which conditions such procedure is successful. Every newly developed model has to be solved individually from the beginning, since there is no unifying solving formalism available.

When shifting the focus towards reaction volumes in the nanometer range, the particle concentration ceases to be a useful concept and has to be abandoned. It is more helpful to focus on the particle number instead. As an alternative to the occupation number representation $p(\mathbf{n}, t)$, one has to focus on distribution functions that specifies both the exact number of particles and their position within the reaction volume. For example, four particles 2A and 2B before a reaction A + B → C could be described by a probability distribution function $P_{AABB}(\mathbf{r}_1, \mathbf{r}_2, \mathbf{r}_3, \mathbf{r}_4, t)$. After the reaction A + B → C has taken place one would have three particles in the system, A, B, and C, and the dynamics of these particles would have to be describe in terms of a new probability distribution $P_{ABC}(\mathbf{r}_1, \mathbf{r}_2, \mathbf{r}_3, t)$. To fully specify the equations that govern the behavior of these probability functions is beyond the scope of this introduction. Equations that interrelate $P_{AABB}(\mathbf{r}_1, \mathbf{r}_2, \mathbf{r}_3, \mathbf{r}_4, t)$ and $P_{ABC}(\mathbf{r}_1, \mathbf{r}_2, \mathbf{r}_3, t)$ are written as

$$\ddot{P}_{AABB}\left(r_1, r_2, r_3, r_4, t\right) = -\left[\sigma_{AB}\left(r_1, r_3\right) + \ldots\right] P_{AABB}\left(r_1, r_2, r_3, r_4, t\right) + \ldots \tag{3.14a}$$

$$\dot{P}_{ABC}\left(r_1, r_2, r_3, r_4, t\right) = -\sigma_{AB}\left(r_1, r_2\right) P_{ABC}\left(r_1, r_2, r_3\right) + \ldots \tag{3.14b}$$

Particular terms were omitted in (3.14). For example, expressions that describe diffusion are absent as well as reaction terms that describe the inflow of probability. Also, not all outflow terms (negative contributions) are included.

It is apparent that when the number of particles increases, the formalism presented in (3.14) becomes impractical both in terms of keeping track of the particles and in terms of performing the actual calculation. For larger numbers of particles, it is more desirable to formulate a theoretical framework that is somewhere in-between the approaches described in (3.11) to (3.13) and (3.14). It is hard to predict where the limit of applicability of (3.14) is, This depends largely on the speed of the computing hardware. An increase in particle number again leads to problems, since the size of the configuration space, after performing a suitable discrimination procedure, scales exponentially with the number of particles.

Additional phenomena have to be considered here. Exclusion effects start to become important, since the size of the reactants is of the same order of magnitude as the size of the reaction volume they are confined in. For example, the presence of a reaction product cannot be strictly neglected anymore: a practice frequently employed when studying infinite volumes. Products of irreversible reactions occupy considerable portions of space and have to be accounted for. Steric effects become important since molecules need to adopt a specific orientation in order to react (Konkoli, Johannesson, et al., 1999). Also, for problems with built-in symmetry (conservation law) one has to keep the information on all distribution functions. For example, in the case of a reaction A + B, the difference in the number of A and B particles is conserved, and higher order distribution functions have to be kept in order to describe the dynamics correctly (Konkoli, Karlsson, et al., 2003). Finally, problems associated with infinite volumes have to be addressed. For example, reactions still influence the pair distribution function which might never reach stationary state, and effective reaction rates become time dependent.

The majority of biochemical reaction processes can be described using the theoretical framework discussed so far (i.e., focusing on DCRs in a range of reaction volumes being extremely large, small, or being somewhere in between). However, when addressing real biological systems at nanometer-length scales, a few additional issues have to be mentioned. As more details are revealed about the complex mechanisms of living cell bioreactions, a picture where reactants move in static, well-isolated volumes becomes less relevant, no matter how large or small that volume may be. Any attempt to successfully model biochemical reactions has to take into account their complexity.

The geometry of the problem can be rather complex. For example, the reaction volume in the cell interior is far from homogeneous or smooth. The concentration of macromolecules in the cytoplasm can be as high as 30% by weight (Luby-Phelps, 2000), resulting in an environment that is inhomogeneous with a high degree of crowding. In vivo reaction volumes are structured, such as in the Golgi apparatus, in mitochondria, or in the endoplasmic reticulum (Stryer, 1992). The geometrical arrangement of reactants can be quite complicated, as in the case of protein complexes that perform metabolic channeling (Kuthan, 2001). It is possible to identify types of chemical reactions that are strongly affected by the fact that geometry is not compact, both in terms of reaction speed and accuracy in generation of the reaction product. There, for instance, it was found that reactions with negative catalytic influences draw benefits from being run in structured space.

When moving from infinite to highly confined spaces, a need for stochastic description becomes apparent. When reaction volumes are small, one has to consider a situation where the number of particles is very low. In principle, all processes on the subcellular level that were discussed in the previous paragraph have to be described in a stochastic manner. However, a full-scale stochastic simulation of such reaction environments is computationally too intensive and highly impractical, if at all possible. In general, it is unlikely that one can describe a highly complicated reaction environment inside the living cell without using some means of multiscale modeling (Takahashi, Yugi, et al., 2002). The associated problem is that special techniques that rely on multiscale modeling remain to be developed, though some work has already been done in this direction. For example, there are studies that focus on separating the deterministic from the stochastic part of the problem (Bentele, Lavrik, et al., 2004) (Bentele and Eils, 2005) (Uhrmacher, Degenring, et al., 2005). The concept here is to use a stochastic description of the reaction pathway, while classical chemical kinetics equations, corresponding to mean field equations for the particle density $\rho(x, t)$, can be used to describe the rest of the problem. A significant challenge in this strategy is to identify the specific parts of chemical reaction pathways that need a stochastic description. Solving this problem could be quite important for reasons other than pure theoretical understanding. For example, in medical biochemistry it might be extremely important to pinpoint which part of a chemical reaction pathway carries or processes the information that is relevant for a specific pathological mechanism leading to disease. In a second stage this knowledge can directly or indirectly lead to detailed understanding of how to manipulate the particular part in order to design a specific drug or treatment. Following a similar line of reasoning, one could address more general questions, such as which processes in the living cell (or larger organisms such as the human body) need to be described using stochastic concepts. One example could be cell apoptosis. Simulations discussed in (Bentele, Lavrik, et al., 2004) revealed interesting features of the threshold mechanism for CD95-induced cell apoptosis that is strongly related to a low (discrete) number of ligands. Moreover, in (Puchalka and Kierzek, 2004) it is shown that stochastic effects arising from dynamics of a few particles can propagate upwards and influence other pathways involving a large number of particles. Very likely, there are many more instances of such behavior that that remain to be discovered.

A multitude of issues needs to be addressed when constructing multiscale models, (e.g., slow and fast degrees of freedom have to be separated). An attempt to develop a formalism to deal with a situation where reaction space is divided in regions of fast and slow diffusion has been presented in (Lizana and Konkoli, 2005). However, systematic understanding of how exactly the dynamics is altered when moving from large, towards medium and small volumes is still lacking (Konkoli, Karlsson, et al., 2003). In general, reaction and transport scales have to be separated.

Both the understanding of fundamental issues and the availability of novel computational tools for modeling diffusion-controlled reactions relevant for living cell biochemistry still need improvement and refinement, especially when treatment of effects of nanoscale systems and volumes is desired. The development of computational techniques that can address such problems is one of the main challenges for future research.

3.4.2 Chemical Transformations in Individual Vesicles

The simplest system to study reactions in a biomimetic membrane environment is by combination of reactants in single phospholipids vesicles. A straightforward fluid membrane reactor concept based on this method has already been reported in 1999 (Chiu, Wilson, et al., 1999). Individual unilamellar phospholipid vesicles with a diameter of a few micrometers were filled with either single reagents or a complete system of reactants and immobilized for manipulation. Mixing and chemical transformations were then initiated either by electroporation or by electrofusion, in each case through application of a short electric pulse, delivered across a pair of microelectrodes. A simple model reaction, converting the dimly fluorescent (background due to unavoidable contamination) chelator calcein into a fluorescent Ca^{2+}-complex was employed, and product formation was monitored by far-field laser fluorescence microscopy. The characteristic of this reaction volume as an ultrasmall confined space led to rapid diffusional mixing and allowed for the study of fast chemical kinetics. This technique has been established as well suited for the study of reaction dynamics of biological molecules within lipid-enclosed nanoenvironments that mimic cell membranes (see Figure 3.11).

Figure 3.10 depicts in two series of images the concept of vesicle-internalized mixing/reaction initiated by electrofusion. In panels 3.10(a–d) a sequence of images shows the loading of adjacently placed phospholipids vesicles with fluorescing dyes of different emission wavelengths, and their combination to a single vesicle by electroporation/electrofusion of the vesicles. The concept has been extended to the initiation and monitoring of a chemical complexation reaction, demonstrating the ability of phospholipids vesicles to function as an ultrasmall volume reactor. In the image sequence 3.10(e–h) the loading of two vesicles with calcein solution (recognizable through its low initial background fluorescence) and a Ca^{2+}-ion solution, and their fusion to a single vesicle. As the main outcome, a brightly fluorescent vesicle interior documents the ability to initiate chemical reactions in a soft-matter container, a starting point for more complex and structured chemical reactor systems, that are not only controllable by membrane fusion, but also by nanotube-mediated merging and by geometrical and dimensional manipulations. The principle has distinct advantages over direct incorporation methods, where the reactants are encapsulated in vesicles during their formation procedure. The low degree of encapsulation, dependence on the

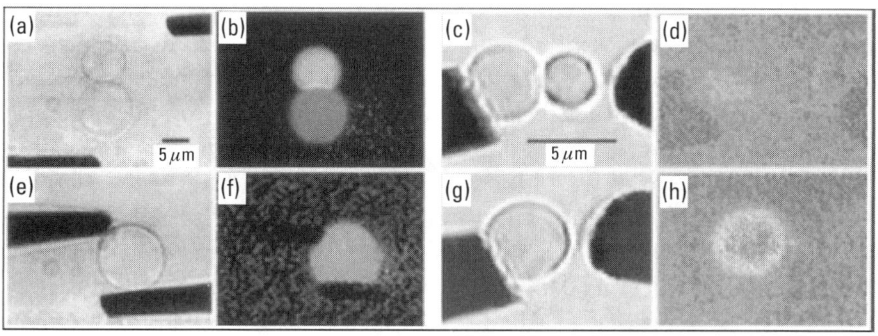

Figure 3.10 Mixing (a–d) and chemical transformation (e–h) in single vesicles initiated by vesicle fusion. (a, c): Brightfield micrograph of a pair of vesicles, each filled with a different fluorescent dye, before (a) and after (c) fusion/mixing. (b, d): laser induced fluorescence images of the same system, before (b) and after (d) fusion/mixing. (e, g) brightfield micrograph of a similar pair of vesicle, one with internalized fluo3, the other one with Ca^{2+}, before (e) and after (g) fusion/reaction, (f, h) Laser induced fluorescence images of the same system, before (f) and after (h) initiation of the reaction. (Images reprinted with permission of *Science Magazine.*)

chemical nature and sensitivity of the material, and concentration limitations are quite effectively overcome.

3.4.3 Enzymatic Reactions in Nanotube-Vesicle Networks

The simple one-container reactor can conveniently be extended to a more structured system, using the soft-matter fabrication technology, outlined in Section 3.2, as a foundation. The NVNs thus obtained can essentially be viewed as biocompatible, transport-controlled reactors that are most advantageously used as hosts for biocatalytic (enzymatic) reactions.

The assumptions of ideal conditions when describing enzymatic reactions in general, such as infinite volumes, homogenous solutions, concentrations of enzymes being much lower than substrate concentrations, and a very dilute solvent system do not apply under conditions present in living cells. Cellular volumes are small and compartmentalized (structured space) and reacting molecules are often associated with membrane surfaces, tubules, or biomacromolecules. Surfaces and microtubules in particular can lead to enhanced reaction rates of diffusion-controlled reactions, since surface-associated molecules need less time to find their reaction partners when the search is limited to less than three dimensions, in contrast to ordinary through-volume-diffusion. Similarly, compartmentalization can enhance reaction rates, since the throughput of substrate and product is more efficient when the participants of a complex reaction are spatially well structured and accessible. Another effect influencing the rates of some reactions or processes, such as protein folding under biological conditions, is macromolecular crowding, arising from the effect of, typically, 5% to 40% protein content in the cellular cytoplasm (Schnell and Turner, 2004). The high concentration of macromolecules leads to an increase in the strength of normally relatively weak molecular interactions, and influences the mobility of affected molecules (Pagliaro, 2000). Moreover, in highly crowded media, the volume for free diffusion of reactants depends on their individual size. Small molecules experience fewer restrictions than larger species, and the rate of reactions involving partners of larger size, being forced to move slower, decreases.

Furthermore, the control of the accessible reaction volume can have a large effect on reaction kinetics. Volume alteration can be achieved by reducing the overall reaction volume, hindering transport, or reducing the dimensionality of the reaction system. In order to investigate chemical and biochemical reactions in restricted environments, a number of solid-state approaches have been developed, such as nanovials (Nagai, Murakami, et al., 2001) (Litborn, Emmer, et al., 1999) or capillaries (Lin, Feldman, et al.,

1997) (Song and Ismagilov, 2003). Apparent problems here are the exposure of reactants, especially biocatalysts, to denaturing, nonbiological surfaces; even more complications arise from evaporation or unspecific adsorption. To circumvent such difficulties, the use of vesicular systems as ultrasmall reactors is sometimes useful; therefore, initiation and control of enzymatic reactions in such biocompatible environments will be discussed in the remaining sections of this chapter.

To illustrate the initiation of chemical reactions in NVNs, an enzyme-catalyzed reaction that produces a fluorescent product was chosen as a model reaction (Karlsson, Sott, et al., 2005). The enzyme alkaline phosphatase, used in the following examples, catalyzes the transformation of its substrate fluorescein diphosphate (FDP) into fluorescein (F) through a two-step cleavage of phosphate groups from the substrate. The (simplified) reaction scheme which applies to this system is schematically depicted in Figure 3.11.

The reaction is fairly complex and many of the microscopic rate constants are not entirely elucidated, yet it can be modeled as a simple first-order reaction during the substrate depletion phase.

3.4.4 Controlled Initiation of Enzymatic Reactions

Even though vesicle fusion and electroporation protocols for reaction initiation are quite effective, all electric field (voltage pulse) based methods suffer from the drawback of release of some material during the time period the membrane needs to react to the pulse. A more considerate way of achieving the same goal of mixing reactant volumes is the merging of vesicles that are interconnected by nanotubes. In this process, the already existing connection between the vesicles is used to avoid the harsh conditions of opening pores in the membrane as a prerequisite for fusion. In principle, several vesicles can be created from a common "mother" vesicle and be temporarily placed on a

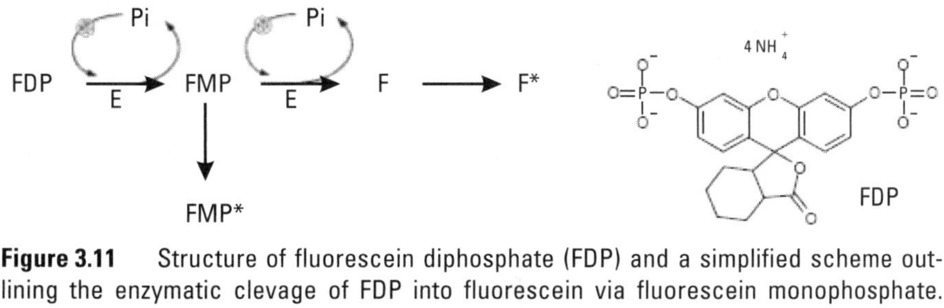

Figure 3.11 Structure of fluorescein diphosphate (FDP) and a simplified scheme outlining the enzymatic clevage of FDP into fluorescein via fluorescein monophosphate. (Reprinted with permission of the American Chemical Society).

surface, while the contents of the injection needle may be changed for every newly created container. Due to the high curvature and correspondingly high state of elastic energy that interconnecting tubes possess, vesicles can be mechanically pushed together, causing the tubes to retract rapidly and to eventually disappear into a joint vesicle with combined volumes. The surplus membrane is drained into the mother vesicle in the process.

Figure 3.12 shows a sequence of images, illustrating this creation and merging procedure. This strategy explicitly avoids the creation of pores in the

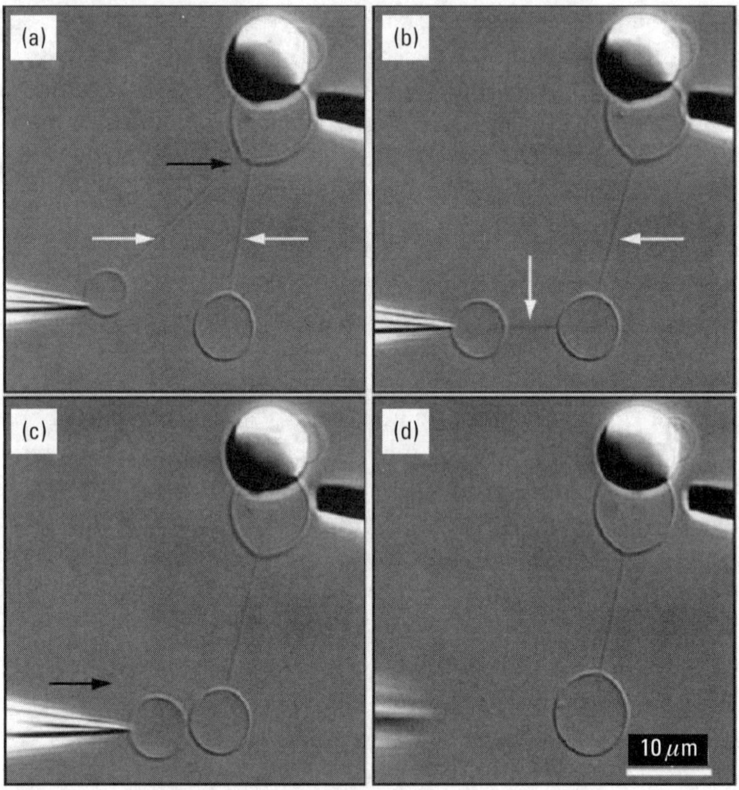

Figure 3.12 Schematic of the nanotube-mediated merging of content-differentiated lipid vesicles. (a, b) Sequential creation of vesicles from a common "mother" vesicle. The black arrow indicates the motion of one nanotube against the other, leading to coalescence. The white arrows indicate the two individual tubes. (c, d) Initiation of a chemical reaction by tube-mediated merging. The black arrow indicated the motion of the vesicle that leads to fusion. Images reprinted with permission of the American Chemical Society.

membrane, and can consequently be used to manipulate very small amounts of enzyme molecules in an accurate manner. In this example, using the alkaline phosphatase/FDP system as a model reaction, product buildup was monitored in a noncontinuous manner. Since considerable photobleaching occurs under laser excitation conditions, a FRAP protocol based on a product buildup/measurement/bleaching cycle was employed for detection (Karlsson, Sott, et al., 2005). The resulting intensity versus time dependency was evaluated, and the rate constant appeared at first sight much lower than that found in bulk measurements of the same enzymatic reaction system. Discrepancies arise from the fact that the enzyme, but not the FDP substrate, is known to adsorb to the walls of the glass injection pipet, lowering significantly the available catalyst concentration. The true number of enzyme molecules in the vesicle was thus determined by comparison with a bulk calibration curve (following pseudo first-order kinetics). This comparison led to the finding that the actual enzyme concentration in the vesicle can be as low as 0.5% to 8% of the initial concentration in the pipet.

The volume of a single reactor of that type was determined to be 1.9×10^{-13} L and the enzyme concentration as 130 pM, which corresponds to 15 enzyme molecules in the interior volume of the vesicle reactor. After depletion of substrate, new vesicles can be filled and the reaction can be reinitiated by the same merging procedure. This technique thus offers the possibility to replenish reacted material and study dilution effects in a confined environment without changing the amount of enzyme involved.

The experimental reaction in this case is too complex to be fully described theoretically, and a simplified, one-step reaction path was used instead of the two-step cleavage of FDP. Based on this simplification, a model was developed and solved with a survival probability approach, assuming an infinite E-S reaction rate. For various reasons, the applied model could not adequately describe the experimental findings quantitatively. However, qualitative trends are sufficiently well pronounced to provide a useful basis for further considerations (Karlsson, Sott, et al., 2005). These findings open pathways to future research and serve as a bridge between simple reactor concepts and complex functional network approaches, as described in the final example.

3.4.5 Control of Enzymatic Reactions by Network Architecture

The proceeding of an enzymatic reaction in a NVN is strongly affected by the geometry and the connectivity of the system. Since the reaction rate is dependent on the diffusional freedom of the system, control and restriction of diffusive transport leads directly to reaction control. A transition from a compact geometry (sphere) to a structured geometry (several spheres connected by nanotubes) in NVNs induces an ordinary enzyme-catalyzed reaction to display wave-like properties. The reaction dynamics are directly and actively controlled by network geometry. Such a networks can be viewed as a chemical waveform synthesizer. The results have bearing both for understanding catalytic reactions in biological systems and for designing soft-matter nanotechnological devices.

An enzymatic reaction, ordinarily following Michaelis-Menten kinetics, can display wave-like behavior where reactions occur as a cascade through a series of reaction nodes being tied together by nanotubes. These systems are continuous with respect to network architecture but initially discontinuous with respect to their interior chemical network connectivity. Network geometry can be controlled dynamically; the flexibility of the lipid material makes it possible to change, for example, network connectivity, tube lengths, or vesicle volumes during the course of the reaction. (Karlsson, Karlsson, et al., 2001) (Karlsson, Sott, et al., 2001) (Karlsson, Sott, et al., 2002) (Karlsson, Sott, et al., 2005) (Sott, Lobovkina, et al., 2006). In this way the network structure can be altered dynamically to amplify and optimize certain properties of the dynamical process it sustains. The catalytic dephosphorylation of FDP by alkaline phosphatase, introduced in the previous section, is used here as a model reaction, and fluorescence arising from the product (fluorescein) is monitored and analyzed in a corresponding kinetic model. Figure 3.13 shows the process of fabricating a chemically differentiated network using the microinjection procedure introduced in Section 3.2. First, the substrate-containing vesicles are created, followed by the enzyme container. A multilamellar vesicle serves as membrane reservoir. The diffusion process, leading to reaction initiation, starts immediately after injecting the enzyme. The total time requirements for this diffusion-controlled reaction (in the order of tens of minutes) makes the short (seconds) injection period uncritical for the course of the reaction.

In this approach, one enzyme-filled vesicle is coupled to several substrate-filled vesicles, and a pattern consisting of waves of product formation develops. Figure 3.14(a–c) shows a linear network where vesicle 1 is filled with enzymes and 2–4 with substrates. Enzyme molecules diffuse through the network and sequentially convert substrates in vesicles 2–4 into products. Figure 3.14(d) shows the corresponding intensity graph. When the catalyst is introduced, the concentration of the different species (i.e., enzyme, substrate, and product) in the downstream containers varies in time as coupled (over-damped) harmonic oscillations where the amplitude and angular

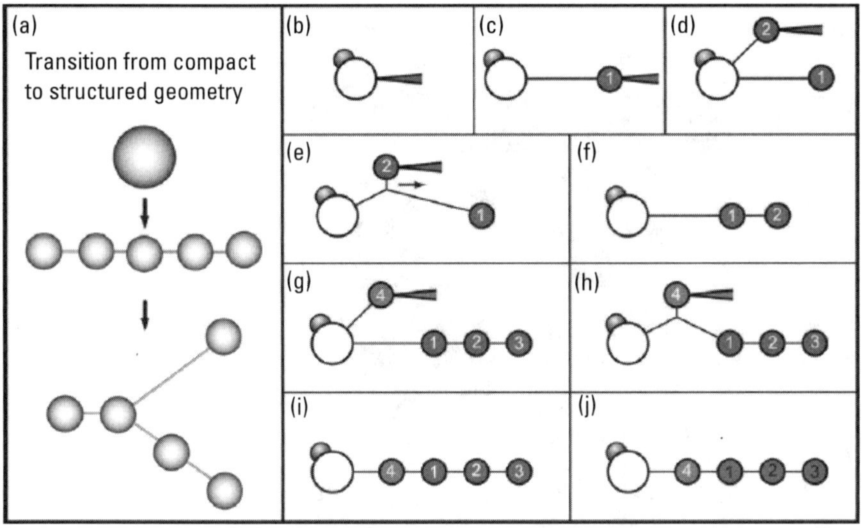

Figure 3.13 Schematic drawings of creation and utilization of the NVN geometry. Panel (a) depicts the sphere to network transition process from compact to structured geometry without changing the topology or volume. Panels (b–j) show the microelectroinjection-based construction of a chemically differentiated network. The numbers of the vesicles reflect the order in which the individual vesicles were generated. Image reprinted with permission of the American Chemical Society.

frequency are functions of network structure and the distribution of reactants. The slope of the intensity curves (i.e., the rate of the reaction) is a function of enzyme concentration (large slope indicates a high enzyme concentration). The height of the curve is proportional to the product concentration and thus the number of available substrate particles in combination with dissipation (i.e., leakage and bleaching). The decrease in fluorescence intensity, after the maximum is reached, is caused by dissipation and substrate depletion.

Transport properties of the reactants vary with network geometry, and therefore the directional, temporal, and spatial coordinates of a reaction wave can be controlled by system configuration parameters. Figure 3.14(e–g) shows a bifurcating network where the substrate vesicles are placed in a V-shaped manner. Initially, vesicle 1 contains enzymes, and vesicles 2–4 contain substrates. Figure 3.14(h) displays the fluorescence intensity measurements for each of the nodes. By diffusion from vesicle 1, the enzyme first reaches vesicle 2 and start to catalyze the conversion of the substrate into the

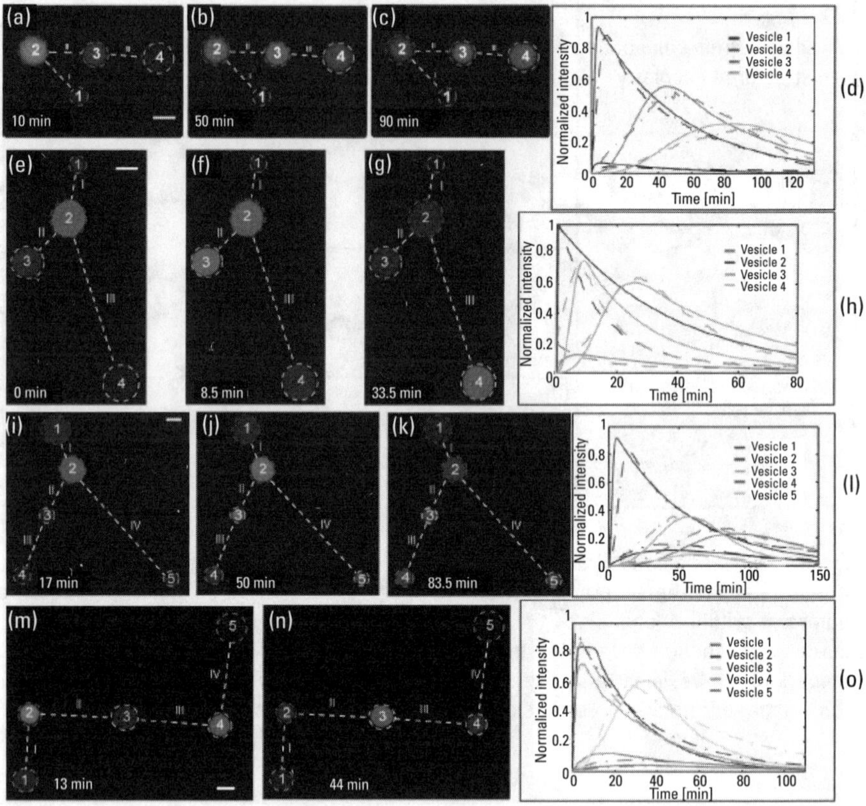

Figure 3.14 Fluorescence microscopy images (a–c, e–g, i–k, m, n) showing product formation in networks with different geometries. (d, h, l, o) show the normalized fluorescence intensities of the corresponding measurements plotted versus time. Initially, the enzyme-filled vesicles are vesicle 1 in panels (a–c), (e–g), and (i–k), and vesicles 1 and 5 in (m–n). The rest of the vesicles are filled with substrate. The dash-dotted lines in graphs (d, h, l, o) show the theoretical fit to the experimentally measured product formation. Fluorescence images were digitally edited to improve image quality. The scale bar represents 10 μm.) Images reprinted with permission of the American Chemical Society.

product. From this bifurcation node, the enzyme diffuses to vesicle 3 and 4 (or returns to vesicle 1). Since nanotube II is shorter (12 μm) than nanotube III (68 μm) they reach vesicle 3 before vesicle 4. Figure 3.14(i–k) shows another bifurcating network similar to the previous except that another vesicle was added to the left branch. Vesicle 1 is filled with enzymes and vesicles 2 to 5 are filled with substrate. Figure 3.14(l) shows fluorescence intensity measurements of product formation in each node. Even though the travel distance between vesicles 2–5 is longer than between vesicles 2–4 (115 μm compared to 79 μm), the enzyme reaches vesicle 5 before vesicle 4. This is because the intermediate container 3 locally dilutes the enzyme concentration and therefore presents a barrier for efficient diffusion of enzyme to the vesicle 4. Thus, the rate-limiting step is not diffusion in the nanotubes (which is fast once the enzymes have entered) but rather the low probability of finding the entrance orifice to the nanotube from the container space (Dagdug, Berezhkovskii, et al., 2003; Lizana and Konkoli, 2005).

The appearances of chemical waveforms evidently depend on connectivity to enzyme-filled vesicles. In addition to the single source of enzyme that diffuses down its concentration gradient in the networks, it can be demonstrated that two discrete sources of enzyme create a pattern of counter-propagating product waves. Figure 3.14(m–o) show product formation in a network consisting of three connected substrate vesicles (vesicles 2–4) positioned between two enzyme-filled containers conjugated to the terminal vesicles 1 and 5. In this system with two sources of counter-propagating enzymes, initial product formation is observed simultaneously in vesicles 2 and 4 (situated close to the enzyme containers), followed in time by product formation in the middle vesicle.

The model developed to fit the experimental data, after application of background correction, is based on rate equations and describes the dynamics of the reaction-diffusion system in NVNs. Reactions in the tubes are neglected due to the very small volume. The rate equations allows for mapping the concentration of enzyme (E), substrate (S), and product (P), respectively, in the different network nodes as a function of time and reads:

$$\partial_t c_{P_j}(t) = \sum_i k_{ij}^{(P)}\left[c_{P_i}(t) - c_{P_j}(t)\right] + k_{cat}\big/K_M c_{E_j}(t) c_{S_j}(t) - k_{dissip.}^{(P)} c_{P_i}(t) \tag{3.15a}$$

$$\partial_t c_{S_j}(t) = \sum_i k_{ij}^{(S)}\left[c_{S_i}(t) - c_{S_j}(t)\right] - k_{cat}\big/K_M c_{E_j}(t) c_{S_j}(t) - k_{dissip.}^{(S)} c_{S_i}(t) \tag{3.15b}$$

$$\partial_t c_{E_j}(t) = \sum_i k_{ij}^{(E)}\left[c_{E_i}(t) - c_{E_j}(t)\right] \tag{3.15c}$$

The three terms on the right-hand side represent transport (diffusion), reaction, and dissipation, respectively. A theory for diffusive transport in a network of containers coupled by thin tubes was recently developed in (Lizana and Konkoli, 2005) where it was demonstrated that the rate for particle transport from container i to j is given by $k_{ij}^{(q)} = D_q \pi a^2 / V_j \ell_{ij}$ where D_q is the diffusion coefficient of substance q, a is the tube radius, ℓ_{ij} is the length of the tube connecting containers i and $j \cdot V_j$ is the volume of vesicle j. A Michaelis-Menten enzymatic reaction is assumed:

$$E + S \underset{k_{-1}}{\overset{k_1}{\rightleftharpoons}} ES \xrightarrow{k_{cat}} E + P$$

and a steady-state approximation on the intermediate enzyme-substrate complex leads to terms containing $c_{E_i}(t)c_{S_i}(t)$ with $K_M = (k_{cat} + k_{-1})/k_1$. In addition to reaction and diffusion, there is continuous loss of particles from the system due to leakage of product through the vesicle wall. The substrate, however, has a low tendency to translocate across bilayer membranes. Additionally, the product is susceptible to photo-bleaching by the high-intensity laser illumination. These leakage and bleaching effects are described by a phenomenological loss term $k_{dissip.}^{(q)}$. The theoretical predictions for diffusing enzyme-catalyzed reactions in the NVNs employed in this study were based on solving (3.15a), which tracks product formation as indicated by dash-dotted lines in the graphs in the Figure 3.14(d, h, l, o).

The behavior of this complex system strongly suggests that the degree of spatial separation of reaction containers in network structures both in biology and in nanofluidic reaction devices can directly influence and modulate the dynamics of an enzymatically catalyzed reaction. Thus, many reactions that appear to be "oscillatory" or time varying, or have otherwise anomalous behavior can sometimes be explained by the particular geometrical structuring of the space in which they occur, that is, they do not need to be inherently autocatalytic or feedback modulated.

3.5 Summary and Outlook

Networks of phospholipids nanotubes and giant unilamellar vesicles are versatile and uniquely flexible architectures for confinement of complex mixtures of chemical reactants and for studies of chemical reactions in ultrasmall volumes, especially in a biologically relevant microenvironment. The advantageous properties of phospholipid bilayer membranes, particularly the bottom-up fabrication strategy by self-assembly from aqueous lipid suspensions and subsequent shape transformations, enable controlled construction of complex, structured container-nanotube assemblies. Several means of membrane functionalization, direct internalization, and manipulation of materials including membrane proteins, enzymes, (bio-)polymers like DNA or poly-(N-isopropyl acrylamide) have been established and evaluated, with the result of a highly versatile biomimetic reactor model. The active and passive transport of small molecules, submicron particles, and biomacromolecules through nanotubes enables initiation and control of chemical reactions within the membrane boundaries of such networks. Due to the small spatial dimensions and short path lengths for molecules to travel, diffusion is the predominant material transport mode, and the impermeability of the phospholipid membrane to all ionic and many nonpolar chemical entities confines reactants to the network interior. Some beneficial and in some cases unique features of vesicle-nanotube reactor networks are summarized in Figure 3.15.

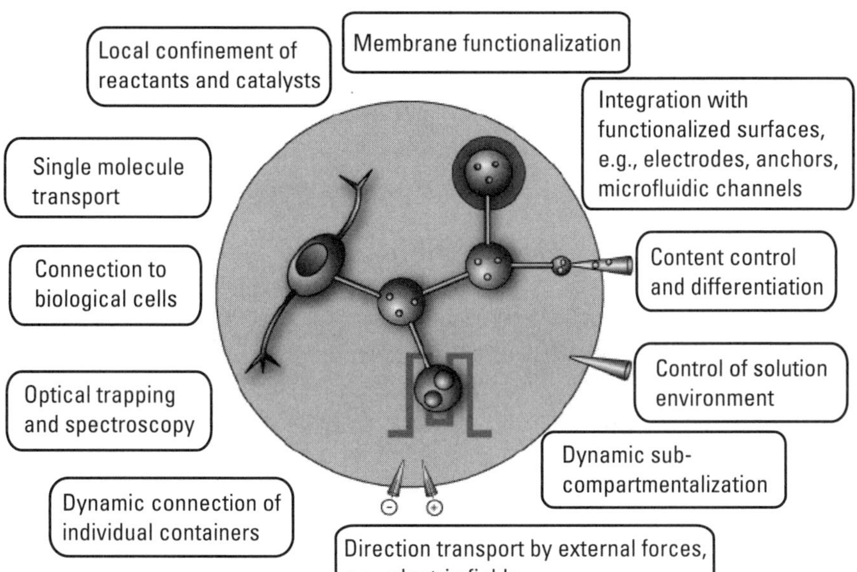

Figure 3.15 Selected unique features of nanotube interconnected vesicle networks composed of soft matter membrane material, providing a multitude of options for initiation and control of chemical reactions, for interfacing to the exterior environment, and for observation and analysis.

Deeper insights into biological and biophysical principles are expected to arise from the development of biomimetic, membrane-based reactor concepts, with high potential to provide better understanding of cellular transport and reaction systems, for example, macromolecular crowding phenomena or the function of natural and artificial ion channels. Entirely new nanoscale fabrication technologies should be expected to emerge, opening doors to functional biomimetic devices of very high complexity. Especially further compartmentalized vesicles with embedded functional units, membrane interacting substructures, and bioelectronic interfaces as well as membrane materials of increased chemical and mechanical stability are some promising directions to follow. Advanced hybrid devices with membrane nanotubes interconnecting vesicles and living cells will eventually lead to new insights into natural transport and reaction phenomena and might aid the extraction and utilization of cell constituents in even more sophisticated biomimetic reactors.

Current limitations of the vesicular reactor networks are, for example, difficulties to extract or remove reaction products from the interior, and to some extent the limited mechanical and chemical stability of the networks due to loss of the membrane integrity over time. These limitations are currently being addressed. Strategies to overcome the issues include chemical stabilization of membranes, hydrogels as interior materials, and more sophisticated "lab-in-a-droplet" setups for precise control of osmotic conditions, surface properties, and pH.

Selected Bibliography

Abdelghani-Jacquin, C., A. Abdelghani, *et al.* (2002). "Decorated surfaces by biofunctionalized gold beads: application to cell adhesion studies." *European Biophysics Journal with Biophysics Letters* 31(2): 102-110.

Alberts, B., A. Johnson, *et al.* (2002). *Molecular Biology of the Cell.* New York, Garland Science.

Angelova, M. I. and D. S. Dimitrov (1986). "Liposome Electroformation." *Faraday Discussions*: 303-308.

Bagatolli, L. A. (2006). "To see or not to see: Lateral organization of biological membranes and fluorescence microscopy." *Biochimica Et Biophysica Acta-Biomembranes* 1758(10): 1541-1556.

Banzhaf, W., P. Dittrich, *et al.* (1996). "Emergent computation by catalytic reactions." *Nanotechnology* 7(4): 307-314.

Barauskas, J., M. Johnsson, *et al.* (2005). "Self-assembled lipid superstructures: Beyond vesicles and liposomes." *Nano Letters* 5(8): 1615-1619.

Bar-Ziv, R., E. Moses, *et al.* (1998). "Dynamic excitations in membranes induced by optical tweezers." *Biophysical Journal* 75(1): 294-320.

Bauer, B., M. Davidson, *et al.* (2006). "Direct reconstitution of plasma membrane lipids and proteins in nanotube-vesicle networks." *Langmuir* 22(22): 9329-9332.

Bentele, M. and R. Eils (2005). "General stochastic hybrid method for the simulation of chemical reaction processes in cells." *Computational Methods in Systems Biology.* 3082: 248-251.

Bentele, M., I. Lavrik, *et al.* (2004). "Mathematical modeling reveals threshold mechanism in CD95-induced apoptosis." *Journal of Cell Biology* 166(6): 839-851.

Bolinger, P. Y., D. Stamou, *et al.* (2004). "Integrated nanoreactor systems: Triggering the release and mixing of compounds inside single vesicles." *Journal of the American Chemical Society* 126(28): 8594-8595.

Borghi, N., O. Rossier, *et al.* (2003). "Hydrodynamic extrusion of tubes from giant vesicles." *Europhysics Letters* 64(6): 837-843.

Brivio, M., W. Verboom, *et al.* (2006). "Miniaturized continuous flow reaction vessels: influence on chemical reactions." *Lab on a Chip* 6(3): 329-344.

Burton, K. and D. L. Taylor (1997). "Traction forces of cytokinesis measured with optically modified elastic substrata." *Nature* 385(6615): 450-454.

Chiu, D. T., C. F. Wilson, *et al.* (1999). "Chemical transformations in individual ultrasmall biomimetic containers." *Science* 283(5409): 1892-1895.

Clark, J., E. M. Singer, *et al.* (2004). "Design and analysis of nanoscale bioassemblies." *Biotechniques* 36(6): 992.

Criado, M. and B. U. Keller (1987). "A membrane fusion strategy for single-channel recordings of membranes usually non-accessible to patch-clamp pipette electrodes." *FEBS Lett* 224(1): 172-6.

Cunningham, C. C. (1995). "Actin Polymerization and Intracellular Solvent Flow in Cell-Surface Blebbing." *Journal of Cell Biology* 129(6): 1589-1599.

Dagdug, L., A. M. Berezhkovskii, *et al.* (2003). "Equilibration in two chambers connected by a capillary." *Journal of Chemical Physics* 119(23): 12473-12478.

Dai, J. W. and M. P. Sheetz (1999). "Membrane tether formation from blebbing cells." *Biophysical Journal* 77(6): 3363-3370.

Darszon, A., C. A. Vandenberg, *et al.* (1980). "Reassembly of Protein-Lipid Complexes into Large Bilayer Vesicles - Perspectives for Membrane Reconstitution." *Proceedings of the National Academy of Sciences of the United States of America-Biological Sciences* 77(1): 239-243.

Davidson, M., P. Dommersnes, *et al.* (2005). "Fluid mixing in growing microscale vesicles conjugated by surfactant nanotubes." *Journal of the American Chemical Society* 127(4): 1251-1257.

Davidson, M., M. Karlsson, *et al.* (2003). "Nanotube-vesicle networks with functionalized membranes and interiors." *Journal of the American Chemical Society* 125(2): 374-378.

Derenyi, I., F. Julicher, *et al.* (2002). "Formation and interaction of membrane tubes (vol 88, art no 238101, 2002)." *Physical Review Letters* 89(20).

Dobereiner, H. G., E. Evans, *et al.* (1997). "Mapping vesicle shapes into the phase diagram: A comparison of experiment and theory." *Physical Review E* 55(4): 4458-4474.

Doeven, M. K., J. H. A. Folgering, *et al.* (2005). "Distribution, lateral mobility and function of membrane proteins incorporated into giant unilamellar vesicles." *Biophysical Journal* 88(2): 1134-1142.

Dommersnes, P. G., O. Orwar, *et al.* (2005). "Marangoni transport in lipid nanotubes." *Europhysics Letters* 70(2): 271-277.

Ellis, R. J. (2001-1). "Macromolecular crowding: an important but neglected aspect of the intracellular environment." *Current Opinion in Structural Biology* 11(1): 114-119.

Ellis, R. J. (2001-2). "Macromolecular crowding: obvious but underappreciated." *Trends in Biochemical Sciences* 26(10): 597-604.

Estes, D. J. and M. Mayer (2005). "Electroformation of giant liposomes from spin-coated films of lipids." *Colloids and Surfaces B-Biointerfaces* 42(2): 115-123.

Evans, E. and D. Needham (1987). "Physical-Properties of Surfactant Bilayer-Membranes - Thermal Transitions, Elasticity, Rigidity, Cohesion, and Colloidal Interactions." *Journal of Physical Chemistry* 91(16): 4219-4228.

Evans, E. and R. Skalak (1980). *Mechanics and Thermodynamics of Biomembranes.* Boca Raton, CRC Press.

Evans, E. and A. Yeung (1994). "Hidden Dynamics in Rapid Changes of Bilayer Shape." *Chemistry and Physics of Lipids* 73(1-2): 39-56.

Eytan, G. D. (1982). "Use of Liposomes for Reconstitution of Biological Functions." *Biochimica Et Biophysica Acta* 694(2): 185-202.

Fishkind, D. J., L. G. Cao, *et al.* (1991). "Microinjection of the Catalytic Fragment of Myosin Light Chain Kinase into Dividing Cells - Effects on Mitosis and Cytokinesis." *Journal of Cell Biology* 114(5): 967-975.

Friedl, P. and K. Wolf (2003). "Tumour-cell invasion and migration: Diversity and escape mechanisms." *Nature Reviews Cancer* 3(5): 362-374.

Fygenson, D. K., J. F. Marko, *et al.* (1997). "Mechanics of microtubule-based membrane extension." *Physical Review Letters* 79(22): 4497-4500.

Ghoroghchian, P. P., P. R. Frail, *et al.* (2005). "Near-infrared-emissive polymersomes: Self-assembled soft matter for in vivo optical imaging." *Proceedings of the National Academy of Sciences of the United States of America* 102(8): 2922-2927.

Girard, P., J. Pecreaux, *et al.* (2004). "A new method for the reconstitution of membrane proteins into giant unilamellar vesicles." *Biophysical Journal* 87(1): 419-429.

Hamley, I. W. (2005). "Nanoshells and nanotubes from block copolymers." *Soft Matter* 1(1): 36-43.

Harris, A. K. (1990). "Protrusive activity of the cell surface and the movements of tissue cells." *Biomechanics of Active Movement and Deformation of Cells.* N. Akkas. 42: 249-294.

Heinrich, V., B. Bozic, *et al.* (1999). "Vesicle deformation by an axial load: From elongated shapes to tethered vesicles." *Biophysical Journal* 76(4): 2056-2071.

Houser, H. (1993). "Phospholipid Vesicles." *Phospholipids Handbook.* G. Cevc. Boca Raton, CRC Press: 603-637.

Huie, J. C. (2003). "Guided molecular self-assembly: a review of recent efforts." *Smart Materials & Structures* 12(2): 264-271.

Hurtig, J., M. Karlsson, *et al.* (2004). "Topographic SU-8 substrates for immobilization of three-dimensional nanotube-vesicle networks." *Langmuir* 20(13): 5637-5641.

Jayaraj, S., S. M. Kang, *et al.* (2007). "A review on the analysis and experiment of fluid flow and mixing in micro-channels." *Journal of Mechanical Science and Technology* 21(3): 536-548.

Jesorka, A., M. Markstrom, *et al.* (2005). "Controlled hydrogel formation in the internal compartment of giant unilamellar vesicles." *Journal of Physical Chemistry B* 109(31): 14759-14763.

Kahya, N., E. I. Pecheur, *et al.* (2001). "Reconstitution of membrane proteins into giant unilamellar vesicles via peptide-induced fusion." *Biophysical Journal* 81(3): 1464-1474.

Karlsson, A., R. Karlsson, *et al.* (2001). "Molecular engineering - Networks of nanotubes and containers." *Nature* 409(6817): 150-152.

Karlsson, A., R. Karlsson, *et al.* (2001). "Networks of nanotubes and containers." *Nature* 409(6817): 150-2.

Karlsson, A., K. Sott, *et al.* (2005). "Controlled initiation of enzymatic reactions in micrometer-sized biomimetic compartments." *Journal of Physical Chemistry B* 109(4): 1609-1617.

Karlsson, M., M. Davidson, *et al.* (2004). "Biomimetic nanoscale reactors and networks." *Annual Review of Physical Chemistry* 55: 613-649.

Karlsson, M., K. Sott, *et al.* (2001). "Micropipet-assisted formation of microscopic networks of unilamellar lipid bilayer nanotubes and containers." *Langmuir* 17(22): 6754-6758.

Karlsson, M., K. Sott, *et al.* (2002). "Formation of geometrically complex lipid nanotube-vesicle networks of higher-order topologies." *Proceedings of the National Academy of Sciences of the United States of America* 99(18): 11573-11578.

Karlsson, R., A. Karlsson, *et al.* (2006). "Chemical analysis in nanoscale surfactant networks." *Analytical Chemistry* 78(17): 5960-5968.

Karlsson, R., A. Karlsson, *et al.* (2003). "A nanofluidic switching device." *Journal of the American Chemical Society* 125(28): 8442-8443.

Karlsson, R., M. Karlsson, *et al.* (2002). "Moving-wall-driven flows in nanofluidic systems." *Langmuir* 18(11): 4186-4190.

Kato, T., N. Mizoshita, *et al.* (2006). "Functional liquid-crystalline assemblies: Self-organized soft materials." *Angewandte Chemie-International Edition* 45(1): 38-68.

Keller, H., P. Rentsch, *et al.* (2002). "Differences in cortical actin structure and dynamics document that different types of blebs are formed by distinct mechanisms." *Experimental Cell Research* 277(2): 161-172.

Kinoshita, T. (1995). "Biomembrane Mimetic Systems." *Progress in Polymer Science* 20(3): 527-583.

Koh, D. S. and D. W. Moon (2006). "Bionanotechnology for single-cell analysis." *International Journal of Nanotechnology* 3(2-3): 314-333.

Konkoli, Z. (2004). "Application of Bogolyubov's theory of weakly nonideal Bose gases to the A+A, A+B, B+B reaction-diffusion system." *Physical Review E* 69(1).

Konkoli, Z. (2005). "Interplay between chemical reactions and transport in structured spaces." *Physical Review E* 72(1).

Konkoli, Z., H. Johannesson, *et al.* (1999). "Fluctuation effects in steric reaction-diffusion systems." *Physical Review E* 59(4): R3787-R3790.

Konkoli, Z., A. Karlsson, *et al.* (2003). "The pair approach applied to kinetics in restricted geometries: Strengths and weaknesses of the method." *Journal of Physical Chemistry B* 107(50): 14077-14086.

Kopelman, R. (1988). "Fractal Reaction-Kinetics." *Science* 241(4873): 1620-1626.

Koster, G., M. VanDuijn, *et al.* (2003). "Membrane tube formation from giant vesicles by dynamic association of motor proteins." *Proceedings of the National Academy of Sciences of the United States of America* 100(26): 15583-15588.

Kotomin, E. and V. Kuzovkov (1992). "Phenomenological Kinetics of Frenkel Defect Recombination and Accumulation in Ionic Solids." *Reports on Progress in Physics* 55(12): 2079-2188.

Kotomin, E. and V. Kuzovkov (1996). *Modern aspects of diffusio-controlled reactions: cooperative phenomena in bimolecular processes.* Amsterdam, Elsevier.

Kuthan, H. (2001). "Self-organisation and orderly processes by individual protein complexes in the bacterial cell." *Progress in Biophysics & Molecular Biology* 75(1-2): 1-17.

Lin, A. L., M. S. Feldman, *et al.* (1997). "Spatially resolved anomalous kinetics of a catalytic reaction: Enzymatic glucose oxidation in capillary spaces." *Journal of Physical Chemistry B* 101(40): 7881-7884.

Litborn, E., A. Emmer, *et al.* (1999). "Chip-based nanovials for tryptic digest and capillary electrophoresis." *Analytica Chimica Acta* 401(1-2): 11-19.

Lizana, L. and Z. Konkoli (2005). "Diffusive transport in networks built of containers and tubes." *Physical Review E* 72(2).

Lobovkina, T., P. Dommersnes, *et al.* (2004). "Mechanical tweezer action by self-tightening knots in surfactant nanotubes." *Proceedings of the National Academy of Sciences of the United States of America* 101(21): 7949-7953.

Long, M. S., C. D. Jones, *et al.* (2005). "Dynamic microcompartmentation in synthetic cells." *Proceedings of the National Academy of Sciences of the United States of America* 102(17): 5920-5925.

Luby-Phelps, K. (2000). "Cytoarchitecture and physical properties of cytoplasm: Volume, viscosity. diffusion, intracellular surface area." *International Review of Cytology - a Survey of Cell Biology*, Vol 192. 192: 189-221.

Luisi, P. L. and P. Walde (1999). *Giant Vesicles*, John Wiley & Sons.

Mallik, R. and S. P. Gross (2004). "Molecular motors: Strategies to get along." *Current Biology* 14(22): R971-R982.

Markstrom, M., A. Gunnarsson, *et al.* (2007). "Dynamic microcompartmentalization of giant unilamellar vesicles by sol—gel transition and temperature induced shrinking/swelling of poly(N-isopropyl acrylamide)." *Soft Matter* 3(5): 587-595.

Mavroidis, C., A. Dubey, *et al.* (2004). "Molecular machines." *Annual Review of Biomedical Engineering* 6: 363-395.

Mills, J. C., N. L. Stone, *et al.* (1998). "Apoptotic membrane blebbing is regulated by myosin light chain phosphorylation." *Journal of Cell Biology* 140(3): 627-636.

Nagai, H., Y. Murakami, *et al.* (2001). "Development of a microchamber array for picoliter PCR." *Analytical Chemistry* 73(5): 1043-1047.

Needham, D. and D. V. Zhelev (1996). "The mechanochemistry of lipid vesicles examined by micropipet manipulation techniques." *Vesicles*. M. Rosoff. New York, Marcel Dekker. 62: 373-443.

Nepomnyashchy, A. A., M. G. Velarde, *et al.* (2002). *Interfacial Phenomena and Convection.* London/Boca Raton, Chapman & Hall/CRC.

Oberholzer, T. and P. L. Luisi (2002). "The use of liposomes for constructing cell models." *Journal of Biological Physics* 28(4): 733-744.

Ovcinnikov, A. A., S. F. Timasev, *et al.* (1989). *Kinetics of diffusion controlled chemical processes*, Nova Science.

Pagliaro, L. (2000). "Mechanisms for cytoplasmic organization: An overview." *International Review of Cytology - a Survey of Cell Biology*, Vol 192. 192: 303-318.

Pitchiaya, S. and Y. Krishnan (2006). "First blueprint, now bricks: DNA as construction material on the nanoscale." *Chemical Society Reviews* 35(11): 1111-1121.

Pohorille, A. and D. Deamer (2002). "Artificial cells: prospects for biotechnology." *Trends in Biotechnology* 20(3): 123-128.

Puchalka, J. and A. M. Kierzek (2004). "Bridging the gap between stochastic and deterministic regimes in the kinetic simulations of the biochemical reaction networks." *Biophysical Journal* 86(3): 1357-1372.

Rentsch, P. S. and H. Keller (2000). "Suction pressure can induce uncoupling of the plasma membrane from cortical actin." *European Journal of Cell Biology* 79(12): 975-981.

Rigaud, J. L., B. Pitard, *et al.* (1995). "Reconstitution of Membrane-Proteins into Liposomes - Application to Energy-Transducing Membrane-Proteins." *Biochimica Et Biophysica Acta-Bioenergetics* 1231(3): 223-246.

Schnell, S. and T. E. Turner (2004). "Reaction kinetics in intracellular environments with macromolecular crowding: simulations and rate laws." *Progress in Biophysics & Molecular Biology* 85(2-3): 235-260.

Schutz, K. and H. Keller (1998). "Protrusion, contraction and segregation of membrane components associated with passive deformation and shape recovery of Walker carcinosarcoma cells." *European Journal of Cell Biology* 77(2): 100-110.

Segota, S. and D. Tezak (2006). "Spontaneous formation of vesicles." *Advances in Colloid and Interface Science* 121(1-3): 51-75.

Seifert, U. (1997). "Configurations of fluid membranes and vesicles." *Advances in Physics* 46(1): 13-137.

Song, H. and R. F. Ismagilov (2003). "Millisecond kinetics on a microfluidic chip using nanoliters of reagents." *Journal of the American Chemical Society* 125(47): 14613-14619.

Sorde, N., G. Das, *et al.* (2003). "Enzyme screening with synthetic multifunctional pores: Focus on biopolymers." *Proceedings of the National Academy of Sciences of the United States of America* 100(21): 11964-11969.

Sott, K., M. Karlsson, *et al.* (2003). "Micropipet writing technique for production of two-dimensional lipid bilayer nanotube-vesicle networks on functionalized and patterned surfaces." *Langmuir* 19(9): 3904-3910.

Sott, K., T. Lobovkina, *et al.* (2006). "Controlling Enzymatic Reactions by Geometry in a Biomimetic Nanoscale Network." *Nano Letters* 6(2): 209-214.

Stryer, L. (1992). *Biochemistry.* New York, W.H. Freeman & Co.

Svenson, S. (2004). "Controlling surfactant self-assembly." *Current Opinion in Colloid & Interface Science* 9(3-4): 201-212.

Takahashi, K., K. Yugi, *et al.* (2002). "Computational challenges in cell simulation: A software engineering approach." *Ieee Intelligent Systems* 17(5): 64-71.

Tanaka, M. and E. Sackmann (2005). "Polymer-supported membranes as models of the cell surface." *Nature* 437(7059): 656-663.

Tank, D. W., E. S. Wu, *et al.* (1982). "Enhanced Molecular Diffusibility in Muscle Membrane Blebs - Release of Lateral Constraints." *Journal of Cell Biology* 92(1): 207-212.

Taylor, T. M., P. M. Davidson, *et al.* (2005). "Liposomal nanocapsules in food science and agriculture." *Critical Reviews in Food Science and Nutrition* 45(7-8): 587-605.

Tien, H. T., Z. Salamon, *et al.* (1991). "Lipid Bilayer-Based Sensors and Biomolecular Electronics." *Critical Reviews in Biomedical Engineering* 18(5): 323-340.

Tokarz, M., B. Akerman, *et al.* (2005). "Single-file electrophoretic transport and counting of individual DNA molecules in surfactant nanotubes." *Proceedings of the National Academy of Sciences of the United States of America* 102(26): 9127-9132.

Trinkaus, J. P. (1973). "Surface-Activity and Locomotion of Fundulus Deep Cells During Blastula and Gastrula Stages." *Developmental Biology* 30(1): 68-103.

Uhrmacher, A. M., D. Degenring, *et al.* (2005). Discrete event multi-level models for systems biology. *Transactions on Computational Systems Biology I*: 66-89.

Weibel, D. B. and G. M. Whitesides (2006). "Applications of microfluidics in chemical biology." *Current Opinion in Chemical Biology* 10(6): 584-591.

Wu, L. Q. and G. F. Payne (2004). "Biofabrication: using biological materials and biocatalysts to construct nanostructured assemblies." *Trends in Biotechnology* 22(11): 593-599.

Yen, A., Y. E. L. Koo, *et al.* (1996). "Experimental study of a crossover from nonclassical to classical chemical kinetics: An elementary and reversible A+B<->C reaction-diffusion process in a capillary." *Physical Review E* 54(3): 2447-2450.

Zana, R. (2005). *Dynamics of Surfactant Self-Assemblies: Micelles, Microemulsions, Vesicles and Lyotropic Phases.* Boca Raton, CRC Press.

Zauner, K. P. (2005). "Molecular information technology." *Critical Reviews in Solid State and Materials Sciences* 30(1): 33-69.

Zhang, Y., F. Gao, *et al.* (2000). "Mechanically gated channel activity in cytoskeleton-deficient plasma membrane blebs and vesicles from Xenopus oocytes." *Journal of Physiology-London* 523(1): 117-130.

4

Ordered Mesoporous Materials

Robert Nooney

4.1 Introduction

The prefix "meso" is derived from the Greek word *mesos* meaning middle and, according to IUPAC classification [1], mesoporous materials have pore diameters ranging from 2 to 50 nm. Materials with diameters smaller that 2 nm and larger than 50 nm are classified as micro- and macroporous, respectively. With regard to the design of biosensors or nanoreactors for applications in life sciences, the most likely application of mesoporous materials would be in the separation or sorting of biomaterials based on size exclusion. This is because mesoporous materials have pore dimensions which match closely with a variety of common proteins. At the start of the 1990s, all hard mesoporous materials were amorphous solids such as silica, activated carbons, transitional aluminas, or modified layered materials, such as pillared clays and silicates. The pores in these structures were irregularly spaced and broadly distributed in size and, despite many efforts, mesoporous materials with regular, well-defined channel systems remained elusive. Nonetheless, amorphous hard materials have been used frequently as a support for the immobilization of biomaterials, particularly vitamins and cellular structures via the sol-gel process. However, the amorphous nature of this material has lead to irreproducible encapsulation and activity.

In 1992, scientists from the Mobil Corporation published the synthesis of a novel mesoporous silicon dioxide powder called MCM-41 (Mobil Composition of Mater-41) [2, 3]. This material contained monodispersed pores surrounded by an amorphous silica wall with long-range order in its pores, which were observable using both TEM and X-ray diffraction. It was the first material to be classified as an ordered mesoporous material (OMM). A transmission electron micrograph of a mesoporous silica nanoparticle with monodispersed pores is shown in Figure 4.1.

Although MCM-41 was classified first, it is not the first occurrence of OMM synthesis. In 1997, Di Renzo et al. repeated a procedure for the synthesis of low-density silica from a patent filed in 1969 by Chiola et al. and obtained a material identical to that published by Mobil [5, 6]. Nonetheless, it is the Mobil scientists who are credited with recognizing the potential of ordered mesoporous materials and who are responsible for igniting scientific interest in this material, which has continued to expand ever since. A search using the online academic database, Web of Science, from 1992 to 2006, returned over 12,300 publications containing the word mesoporous in their title or abstract.

It is expected that OMMs will provide a more controlled environment for the size-selective encapsulation of biomaterials with greater stability and functionality than other amorphous hard materials. There are now many OMMs made from a range of inorganic materials with both controlled pore dimensions and pore symmetry. Moreover, it is now possible to control the overall particle architecture of these materials on the nano- or micrometer scale. For example, several mesoporous silica monoliths have been prepared

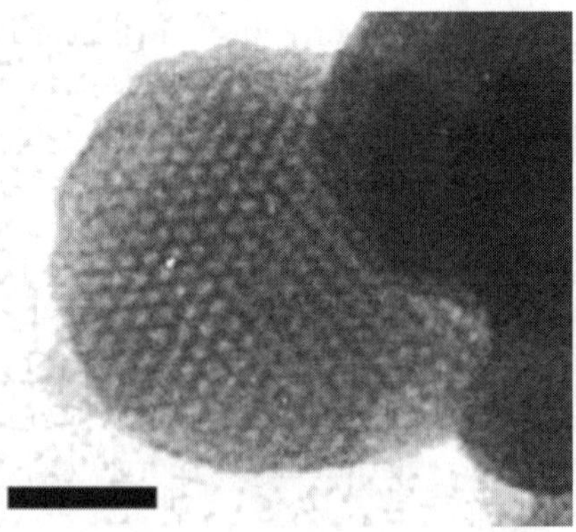

Figure 4.1 A transmission electron micrograph of a mesoporous silica nanoparticle. The scale bar correspond to a length of 25 nm [4].

in a variety of different shapes such as gyroids, helicoids, or discoids [7–9]. These new materials offer great potential in the design of novel micromechanical devices for application in life sciences or medicine.

4.2 The Mechanism of Self-Assembly of Mesoporous Materials

The first OMMs were prepared via the hydrothermal conversion of an inorganic gel, commonly silica, in the presence of a structure-directing template. The template was a surfactant containing a hydrophilic head group and a long carbon tail. In aqueous solutions the carbon tails clustered to form micelles with the head groups facing out into the water phase. The inorganic material was attracted to the hydrophilic head groups and precipitated around the template. The formation mechanism is similar to the formation of a zeolite, whereby charged silica and alumina precipitate around an organic template. Finally, the template was removed via calcination or solvent extraction leaving the inorganic shell (see Figure 4.2 for a computer image of an OMM).

The basic mechanism for the synthesis of OMMs is the liquid crystal templating (LCT) mechanism proposed by Beck et al. (see Figure 4.3) [2]. There are two possible pathways for the self-assembly of OMMs according to Beck's mechanism. The first is a liquid-crystal-initiated pathway whereby the surfactant self-assembles into an ordered material, for example a micellar rod or hexagonal array, prior to the precipitation of the inorganic phase. The second is called the silicate-initiated pathway whereby the silica forms negatively charged oligomers that attract the surfactant molecules and provide the necessary free energy for the self-assembly to proceed.

Mesostructures are unusual, because they can form at pHs above 12 (where silica is soluble) and at surfactant concentrations well below the minimum concentration required for liquid crystal structures to form under aqueous conditions. For example, synthesis of OMMs has been performed using the surfactant cetyltrimethylammonium bromide (CTABr) at concentrations of less than 1 wt %. Normally, aqueous CTAB requires concentrations above 25 wt % to form hexagonal phases and above 70 wt % to form lamellar phases. Clearly there must be a strong interaction between the inorganic gel and the surfactant for the reaction to proceed. This strong reaction is reflected in the rapid precipitation of OMMs when silica precursors and CTAB are mixed at room temperature. Lindel et. al. performed an in-situ X-ray diffraction study of the initial stages of MCM-41 synthesis and observed mesophase formation within the first three minutes of analysis [10].

Figure 4.2 Structure of an OMM with the space group P1 viewed down the (001) axis. The spheres represent oxygen and silica atoms within the amorphous silicon dioxide wall. In this model the pore and wall diameters were set at 3 and 2 nm, respectively.

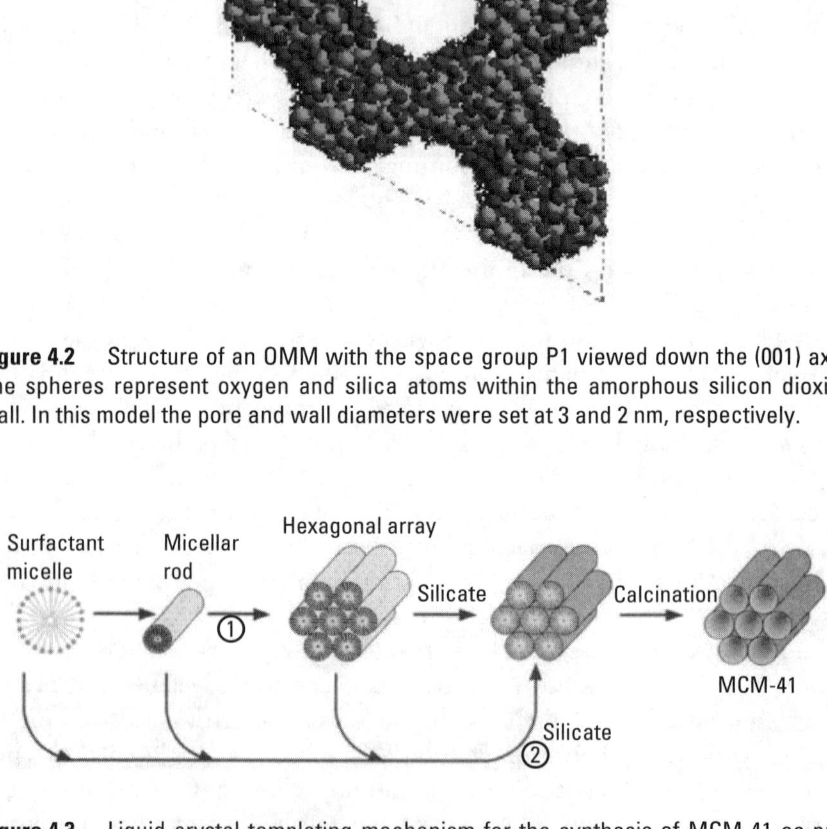

Figure 4.3 Liquid-crystal templating mechanism for the synthesis of MCM-41 as proposed by Beck et al. The first and second pathways relate to the liquid-crystal-initiated and silicate-initiated mechanisms, respectively [2].

Based on an analysis of experimental results, Monnier et al. identified three processes as critical to the formation of a silica mesostructure: 1) multidentate binding of silicate oligomers to the cationic surfactants; 2) preferential silica polymerization in the interface region; and 3) charge density matching between surfactant and the silicate. The Gibbs free energy for the process has been defined in the following way [11].

$$G(A,P) = G\ intra\,(A) + G\ wall\,(P) + G\ inter\,(A,P)\ \text{and}\ G\ sol$$

where G intra (A) relates to the energy inside the surfactant micelle, such as van der Waals interaction potential between each of the carbon chains and electrostatic interactions between the hydrophobic head groups. *G wall (P)* accounts for the internal energy of the silica framework and *G inter (A, P)* defines the energy at the interface between the inorganic phase and the micelle. *G sol* is the free energy of the solution phase. The term *A* refers to the head group area of the surfactant and *P* relates to the composition of the oligomers in the silica wall. Understanding the contributions of each of these parameters to the overall free energy provides a better understanding of the mechanism of OMM formation. The parameters considered to be most important are *G intra (A)* and *G inter (A, P)*. At the start of self-assembly, the silica oligomers are dominated by partially condensed silica groups and have a high surface-charge density. The surfactants are forced to pack closely in an unfavorable formation to balance the charge on the oligomers and lower the free energy of the interface region. At this point, the free energy is dominated by the *G inter (A, P)* term and these OMMs have a lamellar-type symmetry. Mesoporous materials formed at high pHs, where the silica oligomers are highly charged, also exhibit the same free-energy constraints and exhibit lamellar symmetry. On increasing the reaction time, at lower pHs, the silica oligomers condense and lower their surface charge density. The surfactants can now relax to form a mesophase with a lower head group charge density, increasing the free energy of the micelle, *G intra(A)*. This relaxation process leads to a change in the pore structure from a lamellar- to a hexagonal-type symmetry. At this point the structure of the mesophase is controlled more by the *G intra (A)* energy and the head group area, *A*, approaches its optimum value, A_o. One should be aware that optimizing *G intra (A)* and *G inter (A, P)* does not always lead to a simple transformation from a lamellar to a hexagonal symmetry in the mesostructure. There are many other structures that can form including materials with cubic Ia3d and cubic Pm3n symmetries.

Israelachvili et al. stated that it was also possible to predict the structure of a mesoporous material using the dimensionless packing parameter *g*, which is related to A_o by the equation, $g = V/(A_o.l_c)$. *V* is the effective volume of the hydrophobic chain and l_c is the critical hydrophobic chain length of the surfactant [12]. Under circumstances where the surfactant possesses a large polar head group, the packing parameter is small and spherical-type structures are preferred. Alternatively, if the head groups are small and tightly packed, then lamellar or rodlike aggregates are preferred (see Table 4.1).

In a novel study, Huo et al. prepared gemini surfactants, with two quaternary head groups separated by an alkene chain of variable length [13]. This structure is more complex since each surfactant has two charges and the distance between each charge is fixed depending on the alkene chain length. This combination enabled a high degree of control over the size of packing parameter. Using this template, a mesostructure with three-dimensional hexagonal (P63/mmc) symmetry was obtained which has no lyotropic water/surfactant equivalent. This material was labeled SBA-2 after the University of California, Santa Barbara, where it was first synthesized.

There have been many experimental investigations into the effects of varying concentrations of the starting materials on the final structure of the mesoporous material. Vartuli et al. investigated changing the surfactant to silica ratio from 0.5 to 3 and obtained a variety of different structures [14]. There are now several experimentally constructed phase diagrams of the dependence of the mesophase formed on silica precursor, surfactant, and catalyst concentrations available in the literature [15].

It is now possible to make mesoporous materials using a variety of different metal oxides and surfactants under a range of pH conditions. Stucky et al. proposed four general routes for synthesis. The first is the direct condensation of anionic inorganic species with cationic surfactants (denoted S+I– where + denotes the charge on the surfactant, S, and "–" denotes the charge on the inorganic species, I). MCM-41 is a classic example of this; other examples include cubic (point group Ia3d), hexagonal antimony VI oxide, and lamellar tungsten IV oxide [16]. The second route employs a similar

Table 4.1
The Expected Structure of a Mesoporous Material Versus the Dimensionless Packing Parameter, *g*

g	Structure	Examples
1	Lamellar	MCM-41
1/1–2/3	Cubic (Ia3d)	MCM-48
1/2	Hexagonal (p6)	SBA-3
1/3	Cubic (Pm3n)	SBA-1

approach but with the charges reversed (S–I+). Examples include hexagonal phase iron oxide and lamellar phase Mg, Al, Mn, Fe, Co, Ni, and Zn oxides using C12H25PO4H2 as the template [16]. The third route employs a mediated pathway (S+X–I+). The mediating species (X–) is usually a chloride or bromide ion. One example is the synthesis of lamellar zinc phosphate with the composition [CnTMA]+X–[CHZnPO4]+ [16]. The fourth route is similar to the third route with the charges reversed (S–X+I–). One example of this is the synthesis of zinc oxide where CH3(CH2)16COO–X+ is used as the template. Following the four initial routes, a further route involving a neutral templating route was developed by Tanev et al [17]. This material referred to as hexagonal mesoporous silica (HMS), uses a neutral primary amine as the surfactant (So), and a neutral inorganic precursor (Io). The experiment was carried out in deionized water to remove any competing cations or anions that could disrupt the surfactant-inorganic interface. It was assumed the surfactant and inorganic group were attracted to each other via hydrogen bonding. This weak interaction enabled the facile, environmentally benign recovery of costly solvents via solvent extraction methods.

In 1999 Ryoo et al. prepared CMK-1, a new OMM made from carbon (carbon molecular sieve from KAIST, the Korea Advanced Institute of Science and Technology) [18]. In this synthesis, sucrose was absorbed into the pores of calcined mesoporous silica, MCM-48 and converted to carbon through a mild carbonization process using sulfuric acid. The removal of the silica template resulted in a transformation of the carbon structure into a novel, three-dimensional structure with a different symmetry. When SBA-15 was used as the template, an exact replica of the mesoporous silica structure was obtained without any transformation [19]. Unlike hydrophilic silica materials, mesoporous carbon is hydrophobic. This is important because most biological compounds interact more strongly with a hydrophobic surface [20].

4.3 Functionalization of the Pore Walls

Another important feature of all OMMs is the ability to attach functional groups to the pore walls that react strongly with biomaterials. Common functional groups include amines, carboxylic acids, cyanobromide, or sulfhydryls, such as disulfides or maleimides. In the case of silica OMMs these functional groups are attached to the pore wall via the addition of a bifunctional organosiliane. At one end of the organosilane is an alkoxysilane group, such as triethoxysilane for covalent binding to the silica network and at the other end is the bioreactive functional group. A typical example is aminopropyltriethoxysilane. There are two principle mechanisms for attachment of an organosilane: cocondensation and postsynthesis grafting. Cocondensation involves the addition of an organosilane along with the silica source during the self-assembly process. This method has the advantage of distributing the organosilane homogeneously throughout the silica network. However, the organosilane can affect the stability of the silica wall and the energy at the interface between the silica phase and micelle, leading to a less stable mesostructure or a change in symmetry of the material. Postsynthesis grafting requires an additional synthetic step, whereby the organosilane is added after the OMM has been synthesized and generally after the organic template has been removed. However, if the organosilane is too large it can block the pores, reducing the diffusion of biological material into and out of the OMM.

4.4 Controlling the Mesopore Diameter

A unique feature of OMMs is the ability to tailor the pore size by varying the alkyl chain length of the surfactant. Monier et al. reported on the addition of carbons to a quarternary ammonium surfactant $C_nH_{2n+1}[N(CH_3)_3]^+$ for $14 < n < 22$, that the initial pore diameter increased by between 0.1 and 0.12 nm per carbon [11]. Another method for increasing the pore diameters of OMMs is to add auxiliary hydrocarbons such as alkylated benzenes (for example 1, 3, 5-trimethylbenzene) [2]. These smaller organic molecules are called expanders and insert into the micelle arrays causing them to swell in size.

A highly ordered mesoporous material with larger pore diameters and thicker pore walls capable of consuming larger biomolecules was prepared by Zhao et al. [21]. The new material was called SBA-15, (Santa Barbara Amorphous) and was prepared under highly acidic conditions using an amphiphillic triblock copolymer of poly(ethylene oxide) and poly(propylene oxide). The pore diameter ranged from 7.5 to 32 nm with pore wall thickness from 3.1 to 6.4 nm. To achieve the largest pore diameters, the cosolvent, 1,3,5-trimethylbenzene, was added. Different structures were formed depending on the ratio of ethylene oxide to propylene oxide. At low ratios, hexagonal symmetry type materials were formed, whereas at high ratios, cubic type symmetry materials were obtained. Following this work, the same team prepared mesocellular siliceous foam (MSF) using a similar mechanism [22]. This material contained a highly ordered array of large spherical cells connected by narrow windows. This mesostructured material, with its individual cellular compartments, is a promising candidate for the absorption, separation, and transformation of larger biomolecules.

4.5 Characterization

The pore size of OMMs is generally characterized using a combination of liquid nitrogen adsorption, X-ray diffraction, and transmission electron microscopy (TEM). For most OMMs the liquid nitrogen adsorption isotherm is type IV in IUPAC classification [23]. A typical adsorption isotherm of mesoporous nanoparticles is shown in Figure 4.4.

At low-partial pressures, a monolayer of nitrogen molecules is adsorbed onto the mesopore walls and relates to the sharp rise in the amount adsorbed. This is called the Henry's law region where adsorption is proportional to changes in pressure and provides information on the total surface area of the material. Typical OMMs, such as MCM-41, have a surface area approximating $1000 m^2 g^{-1}$, which is greater that the combined area of four tennis courts for one gram of material. The second region of the isotherm, at a partial pressure of approximately 0.2, relates to capillary condensation in the mesopores. A unique feature of isotherms of OMMs is the absence of hysteresis between the adsorption and desorption steps in the capillary condensation region. This is due to the fact that the highly ordered pore structure enables the nitrogen molecules to diffuse into and out of the material without

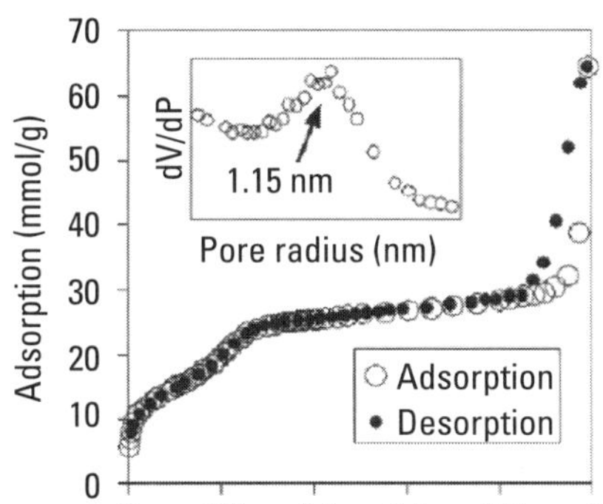

Figure 4.4 Liquid nitrogen adsorption isotherm of mesoporous silica nanoparticles [4]. The inset shows the pore radius distribution determined using the modified Kelvin equation.

obstruction. Less well-ordered materials contain bottlenecks, which lead to higher energy barriers for diffusion of the nitrogen molecules and subsequent hysteresis. The third region at partial pressures close to 1 relates to adsorption between particles and is typical of nanosized mesoporous particles.

The classical method for estimating the pore-size distribution from liquid nitrogen isotherms is to use the Barret-Joyner-Halenda (BJH) modification of the Kelvin equation [23, 24]. The Kelvin equation relates the radius of a pore to the molar volume of the adsorbate and its surface tension. However, it is now thought that the BJH method underestimates mesopore diameters and results from this method should be backed up by other measurements such as TEM. More accurate methods for estimating pore size use either density functional theory, molecular dynamics, or grand canonical Monte Carlo simulations, which model the interaction of the adsorbate with the silica walls at a molecular level [25, 26].

A typical X-ray diffraction pattern of a mesoporous silica nanoparticle is shown in Figure 4.5. The peaks occur at relatively short angles of reflection compared to zeolites and, to a scientist more familiar with the X-ray analysis of zeolites, this result is often surprising. For zeolites, the distance between diffraction planes and, hence, the angle of reflection is related to the distance between the lattice planes of atoms in the crystal, which is of the order of Angstroms. However, for mesoporous materials, the reflections come from Bragg reflections of the replicating silica wall, which is of the order of

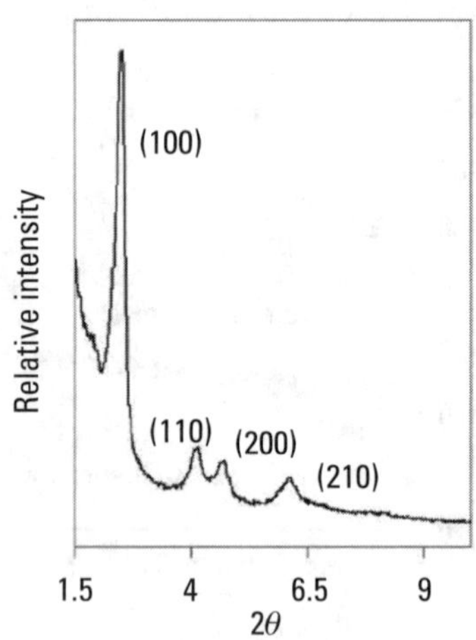

Figure 4.5 An X-ray diffraction pattern of a mesoporous silica nanoparticle. The peaks were indexed to a material with hexagonal pore symmetry. Courtesy of Robert Nooney, University of Notre Dame.

nanometers. The peaks of the MCM-41 powder shown in Figure 4.1 can be indexed to the diffraction planes (100), (110), (200), and (210) of a hexagonal unit cell. The highest number of reflections observed is for mesoporous materials formed on surfaces using evaporation induced self-assembly mechanism under acidic conditions. Such materials are extremely well-ordered over large distances and reflections from the first and second overtones are clearly observed.

Another powerful technique for investigating the pore structure of OMMs is TEM. TEM is an imaging technique whereby a beam of electrons is focused onto a specimen resulting in the capture of an enlarged image on a fluorescent screen, photographic film or CCD camera. A TEM picture of spherical mesoporous NPs with an inset of the honeycomb structure is shown in Figure 4.1. Other features observable using TEM micrographs include lamellar and cubic-type pore structures, disordered regions, and crystal defects, such as disclination planes similar to those found in liquid-crystal-type structures.

4.6 Protein Adsorption and Enzyme Activity

There is a great deal of work on the adsorption and activity of proteins confined in the mesopores of OMMs. The controlled adsorption of proteins is essential in the fields of enzymatic catalysis, biosensors, and disease diagnostics [27, 28].

Vinu et al. studied the adsorption of horse heart cytochrome C and hen-white lysozyme into three different OMMs: MCM-41 (pore diameter 3.54 nm), MCM-41 (pore diameter 4.10 nm), and SBA-15 (pore diameter 10.98 nm) [29, 30]. Horse heart cytochrome C is an electron-carrying protein that switches between ferric (Fe^{3+}) and ferrous (Fe^{2+}) states. It is a spherical protein approximately 3 nm in diameter and is small enough to fit into the pores of all three OMMs. Hen-white lysozyme is a small globular enzyme with two main sections (3.0 by 4.5 nm^2 and 3.0 by 3.0 nm^2) and is also small enough to fit into the pores of all three OMMs. Lysozymes act like antibiotics and bind to the surface of bacteria, facilitating phagocytosis. They are also prevalent in human secretions, including tears and saliva.

Adsorption isotherms of cytochrome C and lysozyme into OMMs at pH 10.5 are shown in Figure 4.6 [29, 30]. The isotherms have type I (or Langmuir) character in IUPAC classification [23]. These isotherms are characteristic of an adsorption process limited by the accessibility of the pore volume rather than the internal surface area. The total amount of protein

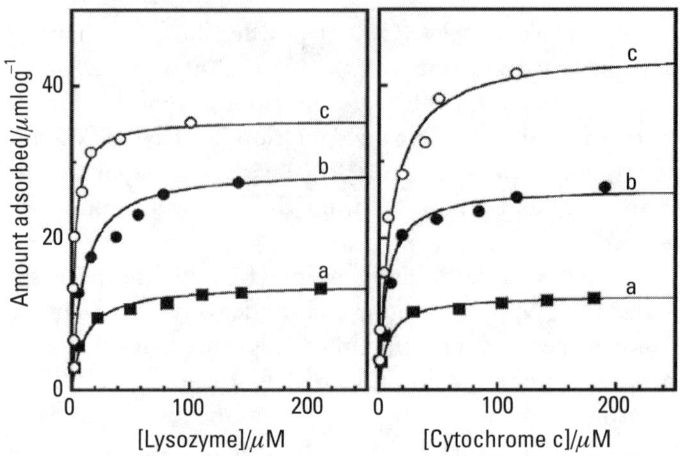

Figure 4.6 Adsorption isotherms of lysozyme and cytochrome in OMM with increasing pore diameter (c<b<a) [29, 30].

adsorbed increased with increasing pore diameter of the OMMs for both proteins. For cytochrome C the partial volume occupied changed from 13.7%, to 26.4%, to 28.3% with increasing pore diameter. For hen-white lysozyme the partial volume occupied changed from 20.1%, to 30.3%, to 35.1% accordingly. Based on mathematical modelling of spherical protein adsorption inside OMMs, these percentages correspond to closely packed configurations of the proteins inside the mesopores [31].

The structural stability of hen-white lysozyme inside the pores of SBA-15 and MCM-41 was investigated using diffuse reflectance infrared Fourier transform analysis. The major absorption bands for lysozyme remained intact on adsorption indicating no denaturation had occurred [29]. Vinu et al. also performed a detailed study on the effects of changing pH on the adsorption of hen-white lysozyme inside OMMs and observed a maximum adsorption at a pH of 10.5 [29]. The isoelectric point of lysozyme is approximately 11 and the isoelectric point of the silica wall is approximately 2.5. If electrostatic interaction between the protein and support is the overriding potential in the absorption process, one would expect an increasing adsorption of hen-white lysozyme as the pH is dropped from 11. However, experimental results showed decreasing adsorption as the pH was dropped. The pH of maximum adsorption is quite close to the isoelectric point of lysozyme at which point the protein is uncharged. This result indicates that the lateral repulsion between adjacent proteins is more significant that the interaction between the proteins and support. Another factor is the diameter of lysozyme, which changes with pH. At a pH of 11 the diameter of lysozyme is 13.5 nm, which is close to its crystalline value. However, at a pH of 4 it is significantly larger, with a diameter of 26.6 nm.

Washman-Kriel and coworkers immobilized several enzymes into OMMs and measured their enzymatic activity [32, 33]. The smaller enzymes, such as cytochrome C and trypsin were found to adsorb to a greater extent than the larger enzyme peroxidase which was assumed to bind only to the surface. Grafting of organosilanes, such as aminopropyltriethoxysilane or 4-trichlorosilyl butyronitrile, stopped the enzymes from leaching at high pH. It was assumed that the organosilanes reduced the pore apertures by up to 1.2 nm, thereby physically entrapping the enzymes. Trypsin is a protease protein and its main biochemical function is to cleave the amide bonds of other proteins. Washman-Kriel et al. measured the hydrolysis of N-á-benzolyl-DL-arginine-4-nitroanilide (BAPNA) using trypsin in MCM-41. Although Trypsin was found to remain stable inside the mesopores, its activity was only 13% of that expected from an equivalent amount of free enzyme. Moreover, it was significantly less active than enzymes bound to other supports such as artificial membranes (activity of 41%), copolymers (65%), and amorphous sol-gel (45%). It was assumed trypsin, being of comparable size to the mesopore, was acting as a plug and hindering diffusion of BAPNA into and out of the pores and that this was the reason for the low activity. Although amorphous hard materials prepared via the sol-gel process display high activity, they have several disadvantages. For example, the enzyme can become denatured during sol-gel synthesis; the enzymes often aggregate and the amorphous pore structure creates obstructions and mass-transfer problems. In a separate study cytochrome C was immobilized inside MCM-48 and its activity was measured using cyclic voltammetry. MCM-48 adsorbed significantly higher amounts of enzyme than MCM-41. This is because it three-dimensional network enabled greater diffusion of the enzyme into the pore network without pore blocking. In this case, the activity was much higher and was comparable to that of enzymes encapsulated in amorphous sol-gel materials. Moreover, MCM-48 provided a model platform with no aggregation, and an open pore structure that could be tailored to match the size of a specific protein. The cytochrome C also retained its redox activity for several months and remained active in aggressive solvents that would normally denature the protein.

Yiu et al. investigated the molecular sieving properties of SBA-15 for a variety of proteins of increasing size [34]. To prevent irreversible adsorption of the protein, a propylthiol group was grafted to the silica wall. The mechanism for adsorption of protein was divided into two steps. The first step was a reversible physisorption interaction between protein and pore wall during which time an equilibrium for the amount adsorbed was established. The second step involved a strong chemisorption between the thiol groups on the pore wall and the protein, leading to an irreversible immobilization of the protein. The thiol functionalized SBA-15 molecular sieve was assumed to have a mean pore diameter of 5.1 nm. The smaller proteins up to and including â-lactoglobulin all showed significant adsorption in SBA-15, but the larger proteins were excluded. This molecular sieving property of OMMs lends itself to possible future applications, such as the separation or sorting of biological material in nanoreactors or micromechanical devices.

A significant problem associated with mesoporous silica materials is their poor stability under aqueous conditions. For materials with thinner pore walls, particularly MCM-41, hydrolysis of siloxane bridges can lead to structural collapse. An alternative material is mesoporous carbon. Unlike silica, it is highly stable under aqueous conditions. Moreover, it is electrically neutral and is less likely to cause deformation of biomolecules immobilized inside its pores. As stated previously, carbon OMMs are prepared via the polymerisation of a carbon source inside a mesoporous silica template. The polymer-silica composite is pyrolyzed at elevated temperatures to carbonize the polymer. Following this, the silica is removed using either concentrated sodium hydroxide or hydrofluoric acid. The adsorption of lysozyme at pH 11 into four carbon OMMs with increasing pore diameters is shown in Figure 4.7 (CMK-1, pore diameter 2.3 nm (a), CMK-3, 3.0 nm (b), CMK-3-130, 4.3 nm (c), and CMK-3-150 6.5 nm (d)) [35]. The carbon OMM with the smallest pore diameter was prepared using MCM-48 as the silica template. The other three materials were prepared using SBA-15 as the template where the pore diameter was increased by raising the temperature of synthesis. The amount adsorbed increased with increasing pore diameter in agreement with published work for adsorption into mesoporous silica materials. The adsorption into CMK-1, which has a pore diameter significantly smaller than the size of the protein is assumed to be adsorption onto the external surface of the material. The total adsorption in CMK-3-150 corresponds to a volume occupancy of only 15%, which is significantly less than the adsorption into silica materials. This low occupancy is likely due to blocking of the pores from crosslinking of the larger carbon rods by small rods creating regions of microporosity. In agreement with work on silica OMMs, the amount adsorbed varied significantly with pH. Again, the maximum adsorption occurred at a pH of 11, close to the isoelectric point of lysozyme. At low and very high pHs, the electrostatic charge on the proteins forced them to move apart leading to much lower packing efficiencies.

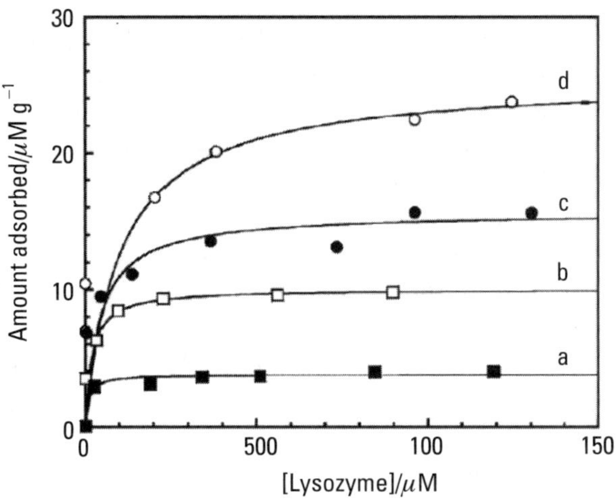

Figure 4.7 Adsorption isotherms of lysozyme into four carbon OMMs with increasing pore size at pH 11. (a) CMK-1, pore diameter 2.3 nm; (b) CMK-3, 3.0 nm; (c) CMK-3-130, 4.3 nm; and (d) CMK-3-150 6.5 nm) [35].

Hartmann et al. studied the adsorption of vitamin E into CMK-3 [36]. Vitamin E is important in the food industry and life sciences. It functions as an antioxidant preventing cell membrane damage and disrupting free-radical chain reactions. Vitamin E, which is hydrophobic, adsorbed more efficiently into the pores of CMK-3 than amorphous activated carbon. The reason for the higher adsorption of vitamin E inside CMK-3 is most likely due to closer contact between the proteins in the ordered pores leading to enhanced van der Waals interactions.

4.7 Morphogenesis of Nano- and Microparticles

So far we have talked about the mechanism and characterization of the pore structure of OMMs, and have not considered ways of controlling the overall size and architecture of the particles. There are now several methods available to control both the size and shape of mesoporous particles on the nano-, micro-, and macrometer scale. Spherical mesoporous particles with sizes ranging from 65 nm to 740 nm can be synthesized using a one-step self-assembly procedure [4]. A silica source, for example tetraethylorthosilicate, is mixed with a surfactant in the presence of a catalyst at low concentrations. The synthesis conditions are called heterogeneous if the silica dissolution time (SDT) is faster than the gelation time (GT), and homogeneous for the reverse case [37, 38]. Heterogeneous synthesis occurs under aqueous conditions and leads to the formation of particles of irregular spherical shape with highly ordered mesoporous channels. Homogeneous synthesis occurs under ethanol/water cosolvent conditions and yields smooth spheres with a starburst mesopore structure extending from the center of the particle to the circumference (Figure 4.8).

The size of the particles can be reduced by increasing the water-to-silica molar ratio, referred to as the R-value in sol-gel chemistry [4]. From transmission electron micrographs in Figure 4.9 we can see that the particle size changes from 70 nm to 460 nmon reducing the R value from 4,000 to 1,000 under heterogeneous conditions. The concentration of the surfactant, CTAB was .7 × 10-3 M for the preparation of sample A, which is significantly lower than previous concentrations used in the synthesis of mesoporous powders (1.1 to 4.9 × 10-2 M) [38, 39]. The concentration of the surfactant is also two orders of magnitude lower than CMC2 for the micelle-to-rod transition under aqueous conditions and further supports the view that particle formation is driven by an interaction between the surfactant and silica as discussed in the formation mechanism section.

Another method for the production of mesoporous spherical nanoparticles is the aerosol assisted self-assembly method, where reactants are confined within an aerosol droplet. As the solvents evaporate, the

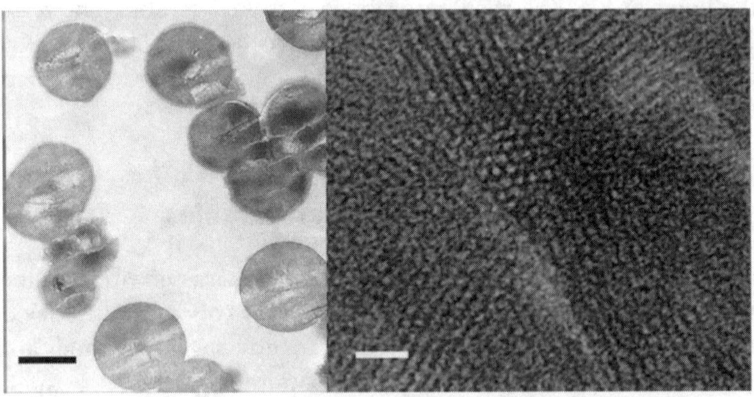

Figure 4.8 Transmission electron micrographs of spherical mesoporous silica nanoparticles prepared under homogeneous conditions (left-hand side with scale bar = 500 nm) and a magnification of the starburst mesopore structure emanating from the center of the nanoparticle (right-hand side with scale bar = 25 nm). Courtesy of Robert Nooney, University of Notre Dame.

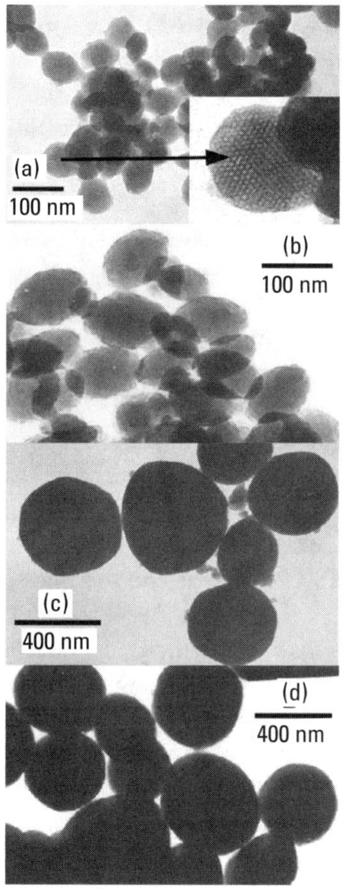

Figure 4.9 Transmission electron micrographs of mesoporous silica nanoparticles prepared under heterogeneous conditions with charged template CTAB. Samples (a–d) correspond to R values or 1,000, 2,000, 3,000, and 4,000, respectively [4].

concentration of the reactants increases to a point where self-assembly of a mesoporous structure is favorable [40].

Simple alkoxysilanes such as tetraethylorthosilicate are not soluble in water but become hydrophilic when partially hydrolysed. Using emulsion chemistry techniques, one can manipulate this changing solubility of alkoxysilanes to synthesize a variety of mesoporous silica morphologies. Schacht et al. added tetraethylorthosilicate (TEOS) to a benzene microemulsion in water stabilized by the surfactant CTAB. The TEOS migrated into the benzene phase where it was soluble and started to hydrolyze. The partially hydrolyzed TEOS moved out of the benzene phase and condensed at the interface region between the water and benzene. The stabilizing CTAB also acted as the template for the formation of a mesophase. This reaction led to the formation of hollow spheres [41]. Alternatively, one can use tetrabutylorthosilicate, TBOS, which forms its own microemulsion in water, negating the need for benzene. As the alkoxysilane hydrolyses it migrates to the interface region and forms a condensed shell (as is the case when using TEOS). Eventually the entire TBOS is consumed and a mesoporous sphere is formed [42, 43].

In the late 1990s, Ozin et al. published a series of papers on the morphogenesis of mesoporous silica under acidic conditions [7–9]. The word "morphogenesis" derives from the Greek words, *morphê* meaning shape and *genesis* meaning creation. Several different architectures were prepared including discoids, flat toroids, spirals, fibers, hollow tubes, and hollow helicoids with dimensions ranging from nanometers to hundreds of microns. In general, synthesis was performed using hydrochloric acid as the catalyst, cetyltrimethylammonium chloride as the template, and TEOS as the silica source. Reactions were performed under quiescent conditions close to the isoelectric point of aqueous silica, which occurs at a pH of 2. High concentrations of ammonium and formate ions were added to enable the formation of mesoporous silica at this low pH.

Morphogenesis was divided into two mechanisms based on the pH of the aqueous solution. At pHs below 2, the reaction was thought to be driven by a silification of a liquid-crystal template, where different shapes originated from disclination defects in the growth of the liquid crystal [8]. Sokolov et al. stated that the free energy of the liquid crystal was inversely proportional to the radius of curvature of the growing mesostructures [44]. At extremely low pHs, (i.e., below the isoelectric point of silica) oligomers are highly charged and would be harder to bend due to electrostatic repulsion, leading to a higher modulus of deformation. To keep the free energy low, the radius of curvature would have to remain high, leading to the formation of straight fibers, as is observed. As the pH is increased to the isoelectric point, the repulsion between neighboring silica groups is reduced thereby enabling the structures to relax to lower energy conformations with higher curvature such as discoids or gyroids (see Figure 4.10). The temperature can also affect the mesostructures observed, because free energy is proportional to thermal energy. Applying the Sokolov relationship, the radius of curvature should also be inversely proportional to the temperature. At lower temperatures, straight fibers are predicted, whereas at higher temperatures, structures with lower radii of curvature, such as discoids and gyroids, are more likely.

Figure 4.10 Scanning electron microscopy images of mesoporous silica particles with gyroid and hellicoid morphologies [7, 9].

Experimental results performed at different temperatures matched these predictions.

The second mechanism is referred to as the supramolecular origami model and occurs at a pH of 1.9, very close to the isoelectric point of silica [9]. In the first stages of this self-assembly process a film forms at the air/water interface. The low acidity and high-ionic strength of the reaction mixture favor a slow rate of silification. As the film thickens and expands across the surface, different rates of polymerization induce contraction of the silicate micelle rods leading to folding of the silica film. Depending on the relative rates of folding parallel and perpendicular to the micelle rods, hollow tubes or helicoids are formed. The hollow helicoids have the shape of an "Archimedian screw" (see Figure 4.10).

Mesoporous silica morphologies may find application in chiral catalysis of macromolecules or the separation of bacterial and viral particles. These architectures could also act as micromolds for the fabrication of metal or magnetic microdevices. The ability to prepare a variety of different structures also points the way to the design of micromechanical machines. The structured materials described above resemble the mineral shells of living organisms such as diatoms and radiolaria. Moreover, the mechanism of morphogenesis may provide a valuable insight into how biomineralization occurs in the natural world.

4.8 Drug Delivery

The ability to load mesoporous materials with biological and chemical compounds make these materials highly desirable as solid phase supports for controlled drug delivery. The general method for drug delivery is to use a support that releases drugs when put in contact with water. For example, biodegradable polymers which release drugs via hydrolysis induced erosion of the carrier structure [45]. However, this method may not be suitable in cases where highly toxic chemicals (such as antitumor drugs) are used and where chemical release before reaching the target cells or tissue cannot be tolerated. If mesoporous materials are used, it is possible to control the time and rate at which the drug is release through controlled blocking of the mesopores.

Kishor, Mal et al. were the first to demonstrate active control over the release of chemicals from an OMM [46]. In this work 7-[(3-triethoxysilyl)propoxyl] coumarin was postsynthetically grafted onto the pore walls of MCM-41. Upon irradiation with UV light of wavelength < 310 nm the coumarin crosslinked to form a cyclobutane coumarin dimer. Subsequent irradiation with UV light, approximately 250 nm in wavelength, photocleaved the coumarin dimers regenerating the coumarin monomers. This process is reversible and the coumarin pore closing and opening action was described as a "double-hinged door." The pores of MCM-41 were filled with the steroid cholestane, which has a molecular diameter small enough to fit inside the pores. The amount of cholestane absorbed, stored, and released was controlled photochemically. The same experiment was repeated using amorphous silica prepared via the sol-gel process. Although photodimerization of the coumarin was still possible in the sol-gel, the cholestane leaked from unblocked pores and it was not possible to control the release rate. MCM-41 was an effective support because the unconnected one-dimensional pores enabled controlled blocking of all the pore openings.

The release of drugs from the mesopores of OMMs can also be controlled chemically. Lai et al. prepared spherical mesoporous nanoparticles approximately 200 nm in diameter and loaded the pores with adenosine triphosphate (ATP) [47]. They then grafted a linear organosilane to the pore walls containing two functional groups; a disulfide bond and a terminal primary amine. Following this, they added a CdS fluorescent quantum dot, QD approximately 2 nm in diameter, functionalized with a carboxylic group. The CdS QD bound to the external pore openings via the formation of an amide bond with the primary amine group of the organosiliane. The CdS QD was very effective in stopping the ATP from leaking from the pores. In water at pH 7.4, only 1% of the ATP, was released over a 12-hour period. However, on addition of a reducing agent, such as dithiothreitol, the disulfide linkage of the organosilane was broken and the CdS QD moved away from the pore opening. In the following 36 hours 85% of the ATP diffused out from the mesoporous nanoparticle.

The biocompatability of this delivery system was tested in vitro using the neuroglial cells, astrocytes. Neuroglial cells provide structural integrity to the nervous system and functional support enabling neurons to transmit information efficiently. The word *neuroglia* in Greek means "nerve glue." ATP binding to the surface of these cells is known to promote the release of [Ca^{2+}], which can be observed using a calcium-binding fluorescent dye. On injection of the mesoporous nanoparticles and subsequent cleavage of the CdS QD, a significant increase in [Ca2+] was observed indicating the ATP had indeed reacted with the astrocyte cells.

In most cases, drug delivery systems have a constant release rate or a rate that decreases with time. However, for many medical conditions such as diabetes or heart rhythm disorders, controlled "on/off" drug delivery in response to an external trigger would be far more desirable. One possible method for on-off or pulsatile drug delivery is the use of ultrasound, which is both noninvasive and can penetrate deep into the interior of the human body. It is also easy to control the physical properties of ultrasound, such as frequency, power density, duty cycle, and time of application. In one study, Kim et al. prepared a composite of a mesoporous silica and poly(dimethylsiloxane) (PDMS) for the controlled release of ibuprofen using an ultrasound trigger [48]. Firstly, mesoporous silica was loaded with ibuprofen and then physically mixed with PDMS. The composite was then cured in a Teflon mold to the desired shape. Figure 4.11 shows the pulsed

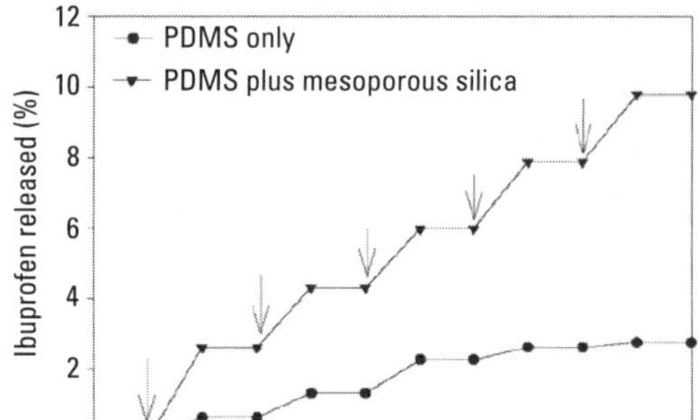

Figure 4.11 Ultrasound-triggered release of ibuprofen from pure PDMS and PDMS-mesoporous silica composite (the down arrows indicate the start of ultrasound irradiation which was continued for a period of 10 minutes) [48].

release of ibuprofen from both a block of pure PDMS and a mesoporous silica PDMS composite using an ultrasound trigger. Although the pure PDMS was loaded with an equal amount of drug the amount released decreased after the second ultrasonic pulse. However, for the composite, the amount released in response to each irradiation cycle was relatively constant.

4.9 Bioactive Glasses for Tissue Engineering

There is a great deal of interest in the use of bioactive glasses (BG) for the reconstruction of bones and teeth. In general, BGs are composed of a combination of SiO_2, CaO, P_2O_5, and MO, where M can be Na or Mg, for example. The BG acts as a glue, enabling implants to bind to soft and hard tissue via the formation of a biologically active hydroxy apatite layer. Hydroxy apatite is a mineral with the chemical formula $Ca_{10}(PO_4)_6(OH)_2$ and is the main mineral component of bone. BGs are also nontoxic and do not induce immune responses or inflammation. In 2004 Yan et al. prepared a mesoporous bioactive glass (MBG) using amphiphillic triblock copolymers as the template [49]. The MBGs were highly ordered with p6mm symmetry, high-surface areas and pore diameters ranging from 4.4 to 7.1 nm in diameter. A standard test of the performance of a BG is to measure the rate of formation of hydroxy apatite on the surface of the BG in the presence of simulated body fluid (SBF) invitro. In general, the rate of formation of hydroxy apatite is extremely slow requiring an induction period of at least 24 hours before any mineral is observed. For MBG, nanoparticles of hydroxy apatite were observed after only 4 hours of soaking in SBF. Moreover, the nanoparticles had a rod-shaped morphology similar to naturally formed particles in human bones. It is likely that the high-surface area and large-pore volume of MBG are the reasons for the increased rate of hydroxy apatite formation. For example, the mesopores could facilitate transportation of dissolved Ca^{2+} and SiO_2 to the mineral as it grows. In further work Shi et. al. mixed MBG with an ammonium phosphate buffer solution to form a paste that could be molded into different shapes or extruded from a syringe [50]. The new material, called mesoporous bioactive glass cement (MBGC), set to form a hard material within 10 minutes. Such a material could be injected into irregularly shaped defects such as cracks in bones or teeth. Moreover, the mesopores can be loaded with drugs or enzymes to accelerate the process of bone formation, or provide nutrients to the bone or scar tissue as it reforms.

4.10 Summary

Ordered mesoporous materials are generally prepared via the hydrothermal conversion of an inorganic gel, commonly silica, in the presence of a structure-directing template. The basic mechanism for the synthesis of OMMs is the liquid-crystal templating (LCT) mechanism. Ultimately, the formation of an OMM requires careful balancing of the Gibbs free energy contributions from the surfactant micelle, the silica wall, and the charged interface region situated between micelle and wall. A simple method for predicting the structure of an OMM is to measure the packing parameter, g. Under circumstances where the surfactant possesses a large polar head group, g is small and spherical type structures are preferred. However, if the head groups are small and tightly packed, g is larger and lamellar or rod-like aggregates are preferred.

It is now possible to prepare mesoporous materials using a variety of different metal oxides and surfactants under a range of pH conditions. Typical metals used include Si, Mg, Al, Mn, Fe, Co, Ni, and Zn. To synthesize carbon OMMs, sucrose is absorbed into the pores of calcined mesoporous silica and converted to carbon through a mild carbonization process using sulfuric acid. The silica is then etched away to leave an ordered carbon network.

The internal pore surface of silica OMMs can be coated with organosilanes for selective binding of biomaterials. At one end of the organosilane is a silicon alkoxide group for covalent binding to the silica network and at the other end is a bioreactive functional group.

The pore size of OMMs can be increased incrementally to match the diameter of a specific protein. The most simple method is to increase the alkyl chain length of the surfactant. If a larger pore size is needed then an auxiliary hydrocarbon is added, for example mesitylene. These hydrocarbons enter the surfactant micelle causing it to swell further. Another strategy for preparing OMMs with very large pore diameters is to use an amphiphillic triblock copolymer as the surfactant. The pore size of OMMs are characterized using a combination of liquid nitrogen adsorption, X-ray diffraction, and TEM.

The measurement of protein adsorption in mesoporous materials is essential in the fields of enzymatic catalysis, biosensors, and disease diagnostics. Proteins small enough to fit inside OMMs show adsorption isotherms with type I character in IUPAC classification. The proteins form a close-packed configuration inside the OMM with partial volume occupancies up to 30.3%. The highest amount of protein adsorbed occurred at a pH close to the isoelectric point of the protein. Larger proteins were excluded from the internal pores and only partly adsorbed onto the external surface of the OMMs. This molecular sieving property of OMMs lends itself to possible future applications, such as the separation or sorting of biological material in nanoreactors or micromechanical devices. Enzymes immobilized in lamellar-type symmetry OMMs exhibited lower activity than enzymes inside amorphous sol-gel. This was because the enzyme acted as a plug and blocked chemical diffusion into the mesopores. OMMs with a three-dimensional mesopore structure, where plugging was not possible, adsorbed significantly more enzyme and showed an activity comparable to amorphous sol-gel. Enzymes immobilized inside OMMs also maintain their activity in aggressive solvents that would normally denature a protein.

Spherical mesoporous NPs can be prepared using a simple self-assembly process where the size of the particles is controlled through a variation of the water-to-silica molar ratio. Other methods for the preparation of spherical mesoporous NPs include the aerosol-assisted method or the formation of a water/oil microemulsion. Under acidic conditions close to the isoelectric point of silica, a variety of OMM with microscale architectures can be prepared. Examples include discoids, flat toroids, fibers, hollow tubes, and hollow helicoids. These architectures are potentially useful in several research areas, such as chiral catalysis of macromolecules, the separation of bacterial and viral particles, the fabrication of micromolds for microdevices, and the construction of micromechanical machines.

OMMs offer several advantages over conventional methods for drug delivery. The common method for delivery is to use a support that releases the drug on contact with water. However, if the drug is highly toxic it is important to control both the start time and release rate. In the case of an OMM fully saturated with a drug, the pore opening can be blocked using either a chemical or a physical barrier. The barrier can then be removed using either chemical or photochemical activation. It is also possible to control the release rate using ultrasound. When the ultrasound is switched on, the drug is released, and when it is switched off delivery stops. This method is also noninvasive and can penetrate deep into human tissue.

Bioactive glasses are used to bind implants to soft and hard tissue via the formation of a mineral layer. Whereas conventional BG takes 24 hours to form a mineral layer mesoporous, BG can form a layer after only four-hours immersion in simulated body fluid. Moreover, mesoporous BG can be mixed with ammonium phosphate buffer solution to form a paste that sets to a hard material in under ten minutes. This paste could be used to repair defects in bones or teeth.

References

[1] IUPAC, "Manual of Symbols and Terminology, Appendix 2, part 1," *Colloids and Surface Chemistry, Pure. Appl. Chem.* 1972, *31*, 578.

[2] J. S. Beck, J. C. Vartuli, W. J. Roth, M. E. Leonowicz, C. T. Kresge, K. D. Schmitt, C. T. W. Chu, D. H. Olson, E. W. Sheppard, S. B. McCullen, J. B. Higgins, J. L. Schlenker, *J. Am. Chem. Soc.* 1992, *114*, 10834.

[3] C. T. Kresge, M. E. Leonowicz, W. J. Roth, J. C. Vartuli, J. S. Beck, *Nature* 1992, *359*, 710-712.

[4] R. I. Nooney, D. Thirunavukkarasu, Y. Chen, R. Josephs, A. E. Ostafin, *Chem. Mater.* 2002, *14*, 4721.

[5] F. DiRenzo, H. Cambon, R. Dutartre, *Microporous Mater.* 1997, *10*, 283,

[6] V. Chiola, J. E. Ritsko, C. D. Vanderpool, *US Patent No 3* 556 725, 1971

[7] G. A. Ozin, H. Yang, I. Sokolov, N. Coombs, *Adv. Mater.* 1997, *9*, 662.

[8] H. Yang. G. A. Ozin, C. T. Kresge, *Adv. Mater.* 1998, *10*, 883.

[9] S. M. Yang, I. Sokolov, N. Coombs, C. T. Kresge, G. A. Ozin, *Adv. Mater.* 1999, *11*, 1427.

[10] M. Linden, S. A. Schunk and F. Schuth, *Angew. Chem. Int. Ed.* 1998, *37*, 821.

[11] A. Monnier, *Science* 1993, *261*, 1299.

[12] J. N. Israelachvili, D. J. Mitchell, B. W. Ninham, *J. Chem. Soc. Faraday Trans.* 1976, *72*, 1525.

[13] Q. Huo, R. Leon, P. M. Petroff, G. D. Stucky, *Science* 1995, *268*, 1324.

[14] J. C. Vartuli, K. D. Schmitt, C. T. Kresge, W. J. Roth, M. E. Leonowicz, S. B. McCullen, S. D. Hellring, J. S. Beck, J. L. Schlenker, D. H. Olson, E. W. J. Sheppard *Chem. Mater.* 1994, *6*, 2317.

[15] G. D. Stucky et al., *Mol. Cryst. Liq. Crsty* 1994, *267*, 1138.

[16] Q. Huo et al., *Nature*, 1994, *368*, 317.

[17] P. T. Tanev , T. J. Pinnavaia, *Science* 1994, *267*, 865.

[18] R. Ryoo, S. H. Joo, S. Jun, *J. Phys. Chem. B*, 1999, *103*, 7743.

[19] S. Jun, S. H. Joo, R. Ryoo, M. Kurk, M. Jaroniec, Z. Liu, T. Ohsuna and O. Terasaki, *J. Am. Chem. Soc.* 2000, 122, 10712.

[20] A. Vinu, M. Miyahara, K. Ariga, *J. Nanosci. Nanotech.* 2006, *6*, 1510.

[21] D. Zhao, J. Feng, Q. Huo, N. Melosh, G. H. Fredrickson, B. F. Chmelka, G. D. Stucky, *Science*, 1998, *279*, 548.

[22] P. Schmidt-Winkel, W. W. Lukens, Jr., D. Zhao, P. Yang, B. F. Chmelka, G. D. Stucky, *J. Am. Chem. Soc.* 1999, *121*, 254.

[23] S. J. Gregg, K. S. W. Sing, *Adsorption, Surface Area and Porosity, 2nd Edition*, Academic Press, London, 1995.

[24] E. P. Barrett, L. G. Joyner, P. P. Halenda, *J. Am. Chem. Soc.* 1951, *73*, 373.

[25] P. I. Ravikovitch, D. Wei, W. T. Chuch, G. L. Haller, A. V. Neimark, *J. Phys. Chem, B.* 1997, *101*, 3671.

[26] M. W. Maddox, J. P. Olivier, K. E. Gubbins, *Langmuir* 1997, *13*, 1737.

[27] B. D. Martins, B. P. Gaber, C. H. Patterson, D. C. Turner, *Langmuir* 1998, *14*, 3971.

[28] T. A. Horbett, J. L. Brash, "Proteins at interfaces II, fundamentals and applications," *ACS*, Washington DC, 1995.

[29] A. Vinu, V. Murugesan, M. Hartmann, *J. Phys. Chem. B.* 2004, *108*, 7323.

[30] A. Vinu, V. Murugesan, O. Tangermann, M. Hartmann, *Chem. Mater.* 2004, *16*, 3056.

[31] M. Miyahara, A Vinu, K. Z. Hassain, T. Nakanishi, K. Ariga, *Thin Solid Films*, 2006, *499*, 13.

[32] J. Felipe Díaz, K. J. Balkus Jr, *J. Mol. Catal. B: Enym* 1996, *2*, 115.

[33] L. Washman-Kriel, V. L. Jimenez, K. J. Balkus Jr., *J. Mol. Catal. B. Enzym* 2000, *10*, 453.

[34] H. H. P. Yiu, C. H. Botting, N. P. Botting, P. A. Wright, *Phys. Chem. Chem. Phys.* 2001, *3*, 2983.

[35] A. Vinu, M. Miyahara, K. Ariga, *J. Phys. Chem. B.* 2005, *109*, 6436.

[36] M. Hartmann, A. Vinu, G. Chandrasekar, *Chem. Mater.* 2005, *17*, 829.

[37] M. T. Anderson, J. E. Martin, J. G. Odinek, P. P. Newcomer, *Chem. Mater.* 2001, *13*, 258.

[38] M. Grun, K. K. Unger, A. Matsumoto, K. Tsutsumi, *Microporous Mesoporous Mater.* 1999, *27*, 207.

[39] Q. Cai, Z. Luo, W. Pang, Y. Fan, X. Chen, F. Cui, *Chem. Mater.* 2001, *13*, 258.

[40] Y. Lu, H. Fan, A. Stump, T. L. Ward, T. Reiker, C. J. Brinker, *Nature*, 1999, *368*, 223.

[41] S. Schacht, Q. Huo,I. G. Voigt-Martin, G. D. Stucky, F. Schuth, *Science* 1996, *273*, 768.

[42] Q. Huo, J. Feng, F. Schuth, G. D. Stucky, *Chem. Mater.* 1997, *9*, 14.

[43] R. I. Nooney, M. Kalyanaraman, G. Kennedy, E. J. Maginn, *Langmuir* 2001, *17*, 528.

[44] I. Sokolov, Y. Kievsky, *Stud. Surf. Sci. Catal.* 2005, *156*, 433.

[45] K. E. Ulrich, S. M. Cannizzaro, R. S. Langer, K. M. Shakesheft, *Chem. Rev.* 1999, *99*, 3181.

[46] N. Kishor Mal, M. Fujiwara, Y. Tanaka, *Nature* 2003, *421*, 350.

[47] C-Y Lai, B. G. Trewyn, D. M. Jeftinija, K. Jeftinija, S. Xu, S. Jeftinija, V. S. –Y. Lin, *J. Am. Chem. Soc.* 2003, *125*, 4451.

[48] H. Kim, H. Matsuda, H. Zhou, I. Honma, *Adv. Mater.* 2006, *18*, 3083.

[49] X. Yan, C. Yu, X. Zhou, J. Tang, D. Zhao, *Angew. Chem. Int. Ed.* 2004, *43*, 5980.

[50] Q. Shi, J. Wang, J. Zhang, J. Fan, G. D. Stucky, *Adv. Mater.* 2006, *18*, 1038.

5

A Novel Nanoreactor for Biosensing

Agnes Ostafin, Hiroshi Mizukami, Philip Wingert, and Hartley Schmidt

5.1 Introduction

Over the past decade, a revolution in biosensing has been underway in response to the rapid advancements taking place in the fields of nanomaterials and nanotechnology. In contrast to the first-generation nanomaterials, which mostly rely on the response of a detectable, nanoparticle-bound molecule to a specific target analyte, these advanced nanodevices are designed to orchestrate interactions among several reactants inside a nano-confinement such that detection is possible only when a target analyte is encountered. Although similar in concept, the complexity of such a material is still quite elementary when compared with what is found in a living cell. Nevertheless, the performance of the reactants in such confinements will be often superior over those implemented as a solution. Presented here is an example how detection of a low concentration of reactive oxygen species (ROS) could be carried out under a hostile environment.

5.2 Basic Design of a Nanoreactor for ROS Detection

5.2.1 Overall Mechanism

The nanoreactor of concern senses reactive oxygen species (ROS) and emits fluorescence at the near-infrared (NIR) region, which is detectable in the biological systems, while being isolated from chemical and spectral interferences (Figure 5.1). Being encapsulated in a biocompatible calcium phosphate cage, the internal reactions are protected from outside contaminants. The ROS permeates through the cage quickly and induces chemiluminescence, whose energy is transmitted via resonance energy transfer (RET) to an acceptor fluorophore and fluoresces at the NIR region. Shifting the emission wavelength to the NIR avoids spectral interferences of the biological material outside the cage and the detection of ROS becomes feasible and simple even in highly absorbing matrices such as blood. Such nanoreactors can be used to evaluate the potential for harm from oxidants, and their quick monitoring may become a vital tool for the prevention of various types of diseases.

5.2.2 Chemiluminescence of Luminol

General mechanisms of luminol chemiluminescence in response to oxidation of peroxide and co-oxidants in alkaline aqueous systems have been proposed previously with hematin as a catalyzer [1–3]. Evidence from the absorption spectropohotometry has suggested that the chemically excited state of 3-aminophthalate (3AP*) ion [Figure 5.2(a)] is the final emitter of chemiluminescence. Intermediate formations of complexes between hematin and peroxide have been suggested, such as a tertiary complex between luminol, hematin, and peroxide, based on the kinetic evidences found for participation of such complexes in the chemiluminescence reaction. According to Olsson [2], the luminol reaction may be formulated for the case of hematin catalysis as shown in Figure 5.2(b).

5.2.3 Resonance Energy Transfer Inside a Nanoreactor

Detection of ROS in biological systems using chemiluminescent light in the range of visible wavelengths is often obscured by endogenous light-absorbing chromophores in the same wavelength range. The motivation to develop a novel ROS sensing nanoreactor suitable in biological systems is dependent on the possibility of avoiding such interferences by shifting the system's light emission to the near-infrared region, taking advantage of RET inside the nanoreactor.

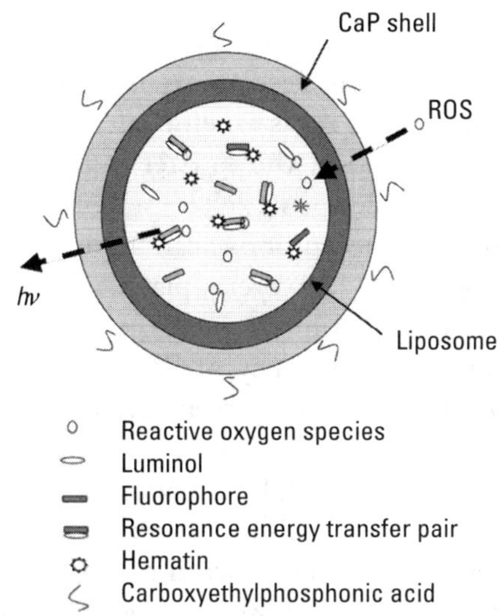

Figure 5.1 Simplified representation of a nanoreactor responding to reactive oxygen species and emitting near-infrared fluorescence.

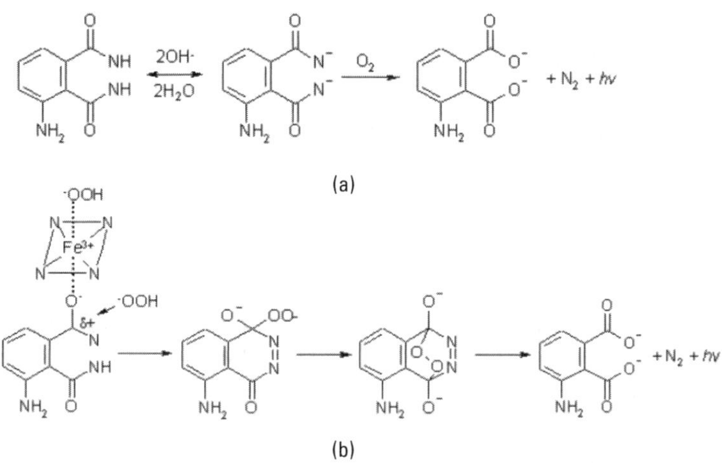

Figure 5.2 (a) Suggested general chemiluminescence mechanism of a luminol-H_2O_2 reaction in basic aqueous solution, where peroxide interacts with luminol in the presence of a catalyst (not shown) to form nitrogen and the excited-state 3-aminophthalate (3AP) ion, which emits light upon its change to ground state. (b) Suggested mechanism of a specific luminol-hematin-H_2O_2 (L-H-P) CL reaction, where it is suggested that an intermediary complex structure between L, H, and P exists as luminol is oxidized by the nucleophilic attack of a second molecule of P. Shown here are the steps from the peroxide adduct to 3-aminophthalate, nitrogen, and light emission.

In solution, RET between an energy donor and acceptor via Förster's dipole–dipole interactions [4] or electron exchange by Dexter's double electron exchange mechanism [5] is inefficient because the chance of donor-acceptor encounter within the Förster radius of acceptor molecule during the lifetime of the donor's excited state is small. Relatively long excited-state lifetimes for the acceptor dye can improve the chances of a fruitful encounter in solution. However, placing these molecules inside nanoreactors where they remain in close proximity with limited mobility is hypothesized to enhance the efficiency of RET between them to levels far greater than that in a solution, leading to emissions mostly at the wavelength of the energy acceptor.

The origins for improved RET in nanoreactors can be most clearly understood by considering the details of Förster's theory which described the rate of nonradiative energy transfer between a stationary excited donor species and a nearby acceptor within a limited range of distance as:

$$k = \left(\frac{1}{\tau_D}\right)\left(\frac{R_0}{r}\right)^6 \tag{5.1}$$

where τ_D is the donor lifetime in the absence of an acceptor, and r is the donor-acceptor distance. The Förster distance R_0 is the distance between molecules at which the energy transfer rate is equal to the decay rate:

$$R_0^6 = 8.785 \times 10^{-5}\left(\frac{\kappa^2 \Phi_D J}{n^4}\right) \tag{5.2}$$

where R_0 is related to the orientation factor, κ^2 between donor and acceptor (generally taken to be an average value of 1/2), the medium's index of refraction, n, Φ_D, the quantum yield of the donor in the absence of acceptor, and J is the overlap integral between the donor and acceptor in the absence of acceptor. It may be expressed as,

$$J = \int F_D(\lambda)\varepsilon_A(\lambda)\lambda^4 \, d\lambda \tag{5.3}$$

In this equation, $F_D(\lambda)$ is the peak-normalized fluorescence spectrum of the donor at wavelength λ and $\varepsilon_A(\lambda)$ is the molar absorptivity of the acceptor.

Despite its widespread use, the orientation factor, κ^2, for dipoles used in Förster's theory is still not a well-understood quantity. In general terms, the longer the lifetime of the *optimal* orientation of two molecules, the faster will be the rate of RET. Förster's theory, originally developed for a stationary donor and acceptor pair, was extended to freely moving molecules in viscous solutions [6, 7] and near surface dipoles [8]. Inside nanoreactors the orientations of charged molecules are significantly restricted by mutual repulsive/attractive interactions with each other, and the nanoreactor walls, so the orientation lifetime of aligned dipoles should be long. Since their translational motions and average separation distance is reduced compared to that in solution, favorable orientations should be maintained even at elevated temperature. These considerations provide an opportunity to design nanoreactors that strongly favor RET between two chromophores inside. Molecular pairs for RET should also have: (1) a sufficient overlap of the emission and excitation spectra of the donor and acceptor; (2) a spatial distance between the two of 1 to 10 nm; (3) a favorable spatial orientation; and if possible; and (4) a donor with a high-fluorescence quantum yield and *long* excited-state lifetime, τ_D.

As mentioned, having a long excited-state lifetime makes a molecule a much better donor, since the efficiency of energy transfer, E_D, is:

$$E_D = 1 - (\tau_{DA} / \tau_D) \tag{5.4}$$

where τ_D is lifetime of the donor, and τ_{DA} lifetime of the donor in the presence of the acceptor [9, 10]. Simply put, there is more opportunity for an energy acceptor to interact with a long-lived donor, which increases the probability that a successful encounter will occur. Confinement of significant numbers of molecule pairs inside a nanoreactor keeps the average distance between them short, reducing the need for a long-lived donor. At the same time, their close proximity increases the likelihood that self-quenching of the terminal fluorescent state will take place. As shown in Figure 5.1 for a nanoreactor containing luminol and a fluorophore, this leads to the appearance of a significant amount of fluorescence emission. In this case, the fluorophore is a good energy acceptor species for luminol but its other characteristics and concentration in this example have not been optimized for the process. The trend however is clear.

In addition to the improved advantage of RET in the nanoenvironment, advantage over collision frequency may also be anticipated. The frequency of collisions, z, between two molecules of diameter, d, in gas is expressed as,

$$Z = 2^{1/2} \pi N d^2 c \tag{5.5}$$

where N is the number of molecules per unit volume and c is the average speed of the molecules. This equation, with some modification, is also applicable to a solution and suggests that the collision frequency increases as the concentration of molecules and/or their size are increased. A higher collision frequency between donor and acceptor dyes will improve the resonance energy transfer efficiency, while a too high frequency could begin to degrade both the stability of the chemically excited luminol intermediate, as well as lead to fluorescence self-quenching between the terminally excited fluorescent molecules.

5.2.4 A Kinetics Model of Nanoreactor Chemiluminescence and Fluorescence

A simplified kinetic mechanism for oxidant detection through chemiluminescence-RET-fluorescence of a nanoreactor is shown in Figure 5.3. In this model, only the major rate constants are indicated. A dotted rectangle represents a nanoreactor and shows that ROS, hydrogen peroxide, enters through the calcium phosphate shell and liposome layer at the permeation rate constant of k_1 and in reverse k_2. Inside the nanoreactor, chemical reaction between hydrogen peroxide and luminol with the help of the molecular catalyst hematin forms an excited state of the 3 aminophthalate (luminol) dianion, 3AP*, at a rate of k_3. A fraction of 3AP* may chemiluminesce at 430 nm and return to the ground state. The intensity of this chemiluminescence is greatly affected by the concentration of some metal ions and hematin. Another fraction of the excited-state 3AP* may experience dipole-dipole interaction with nearby fluorescent molecules, F,

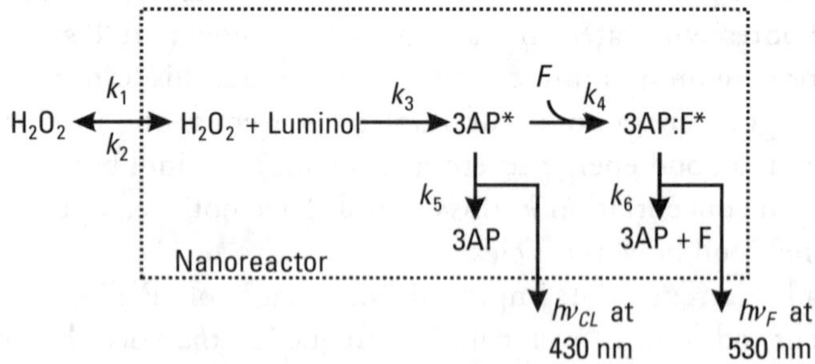

Figure 5.3 Illustration of kinetic steps for ROS detection by a nanoreactor through chemiluminescence resonance energy transfer. Note that both chemiluminescence (430 nm) and fluorescence (530 nm) may be observed.

such as fluorescein, at a rate of k_4. The product of this is an interaction pair 3AP:F*, and, if their interactions satisfy the conditions described for RET, there will be fluorescence from an acceptor fluorophore, $h\nu_{Fl}$, at 530 nm for fluorescein, for example. This simplified figure shows only the emissive pathway for fluorescein relaxation, but some energy will be lost by direct transition of 3AP*, 3AP:F*, and F* to their respective ground states. These processes would become more significant at higher overall concentrations.

The processes shown in Figure 5.3 may be expressed mathematically. First, for the entrance of ROS to the nanoreactor is expressed as:

$$\frac{d[\mathrm{H_2O_2}]_{\mathrm{nanoCRET}}}{dt} = k_1[\mathrm{H_2O_2}]_{\mathrm{outside}} - k_2[\mathrm{H_2O_2}]_{\mathrm{nanoCRET}} - k_3[L][\mathrm{H_2O_2}]_{\mathrm{nanoCRET}} \tag{5.6}$$

and the change of 3AP* may be expressed as,

$$\frac{d[\mathrm{3AP}^*]}{dt} = k_3[\mathrm{H_2O_2}]_{\mathrm{nanoCRET}} - K_{FR}k_4[F][\mathrm{3AP}^*] - k_5[\mathrm{3AP}^*] \tag{5.7}$$

where K_{FR} is a complex value that depends on the condition of the RET pair formation, a given total concentration of fluorescein, as well as that of luminol. Since $\frac{d[3AP^*]}{dt}$ relates to the total intensity of luminescence, it is evident that in the absence of F, the luminescence is only from chemiluminescence. Increasing concentration of F shifts the luminescence to fluorescence, but the increase will not be a linear dependence on the concentration of F and requires experimental determination.

From this simple model it is apparent that there will be a concentration optimum where the detectable, oxidant-dependent fluorescence is at an optimum. At first, in order to confirm the improved resonance transfer in nanoreactor, much higher concentrations of ROS over luminol should be used to keep the level of excited-state intermediate relatively constant. In practice, the nanoshell system is intended to stabilize and maximize the oxidant-dependent signal when the concentration of the oxidant is very small, so oxidative damage to either the luminol or fluorescein is expected to be small during the duty cycle of these materials.

A number of fluorescence dyes, such as fluorescein, eosin Y, and rhodamine 6G are good candidates as the RET acceptors of 3AP* having their absorption maxima at 490, 520, and 525 nm and their dominant emission peaks at 530, 550, and 555 nm, respectively. However, more satisfactory energy acceptor candidate in the biological tissue samples will be water soluble $Ru(bpy)_2(dcpby)(PF_6)_2$ (molecular probes), since its absorption is at 458 nm and emission at 650 nm, which is in the range of NIR.

5.3 Synthesis of a Nanoreactor

5.3.1 Outline of Nanoreactor Synthesis

The steps of nanoreactor synthesis are schematically represented in Figure 5.4. First, the desired kind of phospholipid will be hydrated in a solution of reactants to be encapsulated and multilamellar phospholipid bilayers will be

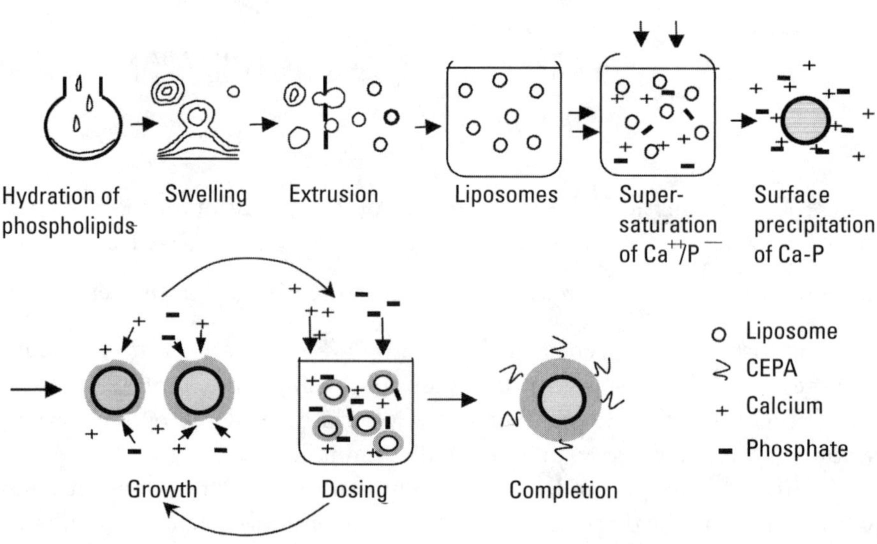

Figure 5.4 A simplified representation of nanoreactor synthesis. The selected phospholipid will be hydrated in a solution of reactants to be encapsulated and multilamellar lipid bilayers will be formed. They are extruded through the membrane having pores of known size to produce uniform liposomes. To the sized liposomes a supersaturated phosphoric acid and calcium chloride solutions are added to initiate precipitation of calcium and phosphate ions on the negatively charged surface of liposomes. Ionic interaction-induced layering of calcium and phosphate ions will continue to increase the thickness of the calcium-phosphate shell as needed, and the growth is terminated by covering the surface with CEPA.

formed. They will be extruded through the membrane having pores of selected size to produce uniform liposomes. To the sized liposomes, a high concentration of phosphoric acid and calcium chloride solutions are added to initiate its supersaturation precipitation [11] on the negatively charged surface of liposomes. Ionic interactions induced layering of calcium and phosphate ions that will continue to increase the thickness of the calcium-phosphate shell as needed, and the growth is terminated by covering the surface with carboxyethylphosphonic acid (CEPA). More detailed description of synthetic procedures will follow.

5.3.2 Encapsulation of the Reactants in Liposomes

Encapsulation process is the critical first step to confine the reactants in a nano-sized liposome compartment, while assuring selected types of ROS to permeate through the liposome layer. Various kinds of phospholipids may be used to synthesize a liposome, but one type of phosphatidic acid lipid, 1,2-dioleoyl-*sn*-glycero-3-phosphate (DOPA), is cited as an example.

Materials: All chemicals used were reagent grade and used without further purification, unless stated otherwise. Hematin (porcine), fluorescein, and carboxyethyl phosphonic acid (CEPA) were obtained from Sigma-Aldrich (St. Louis, Missouri). The phosphatidic acid lipid DOPA was obtained as lyophilized powder from Avanti Lipids (Alabaster, Alabama).

Liposome Preparation: The liposomal templates for aqueous-cored calcium phosphate nanoreactors were formed by the interaction of highly concentrated chemiluminescence reagent stocks containing luminol and hematin close to their maximum solubility levels in phosphate buffered saline (PBS) at pH 9.5. After mixing, the concentration of luminol (L) was approximately 1.0 mM, fluorescein (F) 0.8 mM, and hematin (H) 0.032 mM. An aliquot of the mixture was mixed with layered dry DOPA lipid causing its swelling and budding of the film into multilamellar (onion-like) liposomes.

Liposome Extrusion: Hydrated DOPA lipid suspended in the solution of CL reagents was mixed with a 3/4 inch magnetic stirring bar at 1,000 RPM for 1 hour in a 20-ml scintillation vial. A 25-mm polycarbonate extrusion membrane with 100-nm pore size was fitted to a handheld 1-ml extrusion apparatus (Avanti Polar Lipids). The 100-nm pore-size filter was chosen based on the successful results of previous work [12] and later other pore-size filters were tested for this research to study the relation between the size of a nanoshell and the CL performance. A 1-ml suspension of DOPA lipid-hydrated L-H-F suspension was passed through the extruder 25 times. To measure the particle size distribution, undiluted nanoparticle suspensions were first filtered through a 1.2-μ CA filter to remove dust. A Brookhaven Instruments Corporation (BIC) ZetaPALS Particle Sizer with a He-Ne laser and a detector angle of 90 degrees was used. Each 3-minute scan was averaged 6 times to obtain a standard deviation of the distribution mean size. The results are shown in Figure 5.5 (open circles) and are 100 ± 40 nm in diameter. Once the extrusion was completed, it was observed that there was negligible volume loss due to the filter or inside of the extruder via measuring the loss. The resulting CL liposome suspension was left to rest for 1 hour afterward before coating the liposome with a calcium phosphate shell.

5.3.3 Self-Assembly of Calcium Phosphate Shells over the Liposomes and Nanoreactor Stabilization with CEPA

In contrast to the drop-wise self-assembly of a calcium phosphate shell over the liposomes, an alternative one-step self-assembly under super-saturation appears simple and equally satisfactory results are obtained. A simplified diagrammatic description of super-saturation self-assembly of calcium phosphate shells is already illustrated in Figure 5.4. After CL liposomes were left undisturbed in darkness for 1 hour, 1 ml of the suspension was added to a

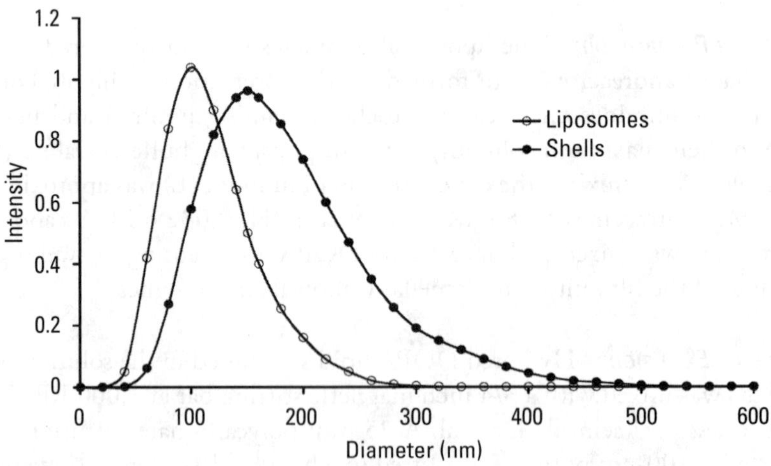

Figure 5.5 Size analysis of liposomes (open circles) and shelled nanoreactors (filled circles) with a ZetaPALS Particle Sizer.

50-ml glass beaker containing 25-ml DI water and 125 ml of 0.1M phosphoric acid adjusted to pH 7.0 with 1M NaOH. Within 5 seconds a 125 ml of 0.1M calcium chloride was added to the solution. The solution was then stirred continuously in a 50-ml glass beaker at 600 RPM for 4 hours. In the basic solution, the surface of DOPA liposomes will be negatively charged and function as the nucleation sites for positive ions. Thus, addition of near-saturated calcium chloride and phosphoric acid solutions will initiate an ionically attracted layer of calcium ions on the surface of liposomes, and in turn the phosphate ions. As will be shown later, they are not in perfect order, but subquent formations of Stern layers will force an increase in the thickness of the now calcium phosphate (Ca/P) shells.

To arrest the nanoshell growth, a 400 ml of 0.1 M CEPA at pH 7 was added and the suspension stirred for an additional 30 minutes. It was then left undisturbed for at least 1 hour. An example of the size distribution of Ca/P shell-coated liposomes (i.e., nanoreactor), is shown in Figure 5.5 (closed circles) and is 150 ± 60 nm in diameter.

5.4 Characterization of a Synthesized Nanoreactor

5.4.1 Physical Feature of a Nanoreactor

A TEM image of a typical nanoreactor is shown in Figure 5.6(a). Shells produced via this synthesis appear hollow and visually more uniform than the

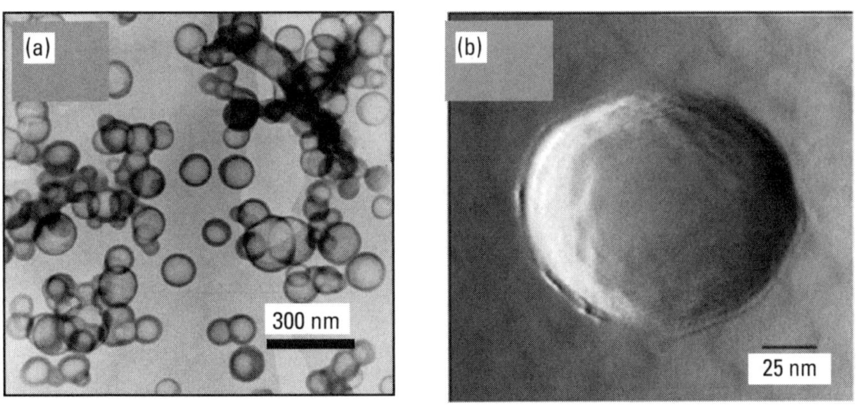

Figure 5.6 (a) A transmission electronmicrograph of a nanoparticles on the left, and (b) AFM images of the same showing a surface curvature on the right.

light-scattering results suggest, with a diameter of 150 ± 60 nm (n = 100) and shell thickness of 10 to 20 nm (n = 100). The diameter of a nanoreactor observed using the atomic force microscopy (AFM) seems to agree with the result of TEM [Figure 5.6(b)]. AFM images of the as-synthesized calcium phosphate nanoreactors were obtained using a Multimode IIIa NanoScope (Digital Instruments) operating in tapping mode with OTESPA probes (Digital Instruments) of 15-nm nominal diameter. The same grid used for visualization by TEM was also used for topographic imaging by AFM. TEM grids were fixed to an AFM sample puck (Ted Pella) with double-stick tape. The NanoScope v. 5.12 r3 software bearing analysis tool was used to determine the average height of the particles. There were between 100 to 1,000 particles per square micron.

The effectiveness of using CEPA as the stabilizer of nanoreactor surface is evident by the increased negativity of the Zeta potential obtained, and as shown in Figure 5.7. The plateau reached indicates that CEPA has saturated the surface and no further increase can be achieved. Zeta potentials were measured by immersing the polycarbonate electrode of the ZetaPALS machine in a 4-ml 1-cm polyacrylate cuvette with 1.5-ml sampler volume. An average of 10 runs was recorded as the mean value with standard deviation in mV.

These observations seem to confirm the assumed structure of a nanoreactor that is shown in Figure 5.1.

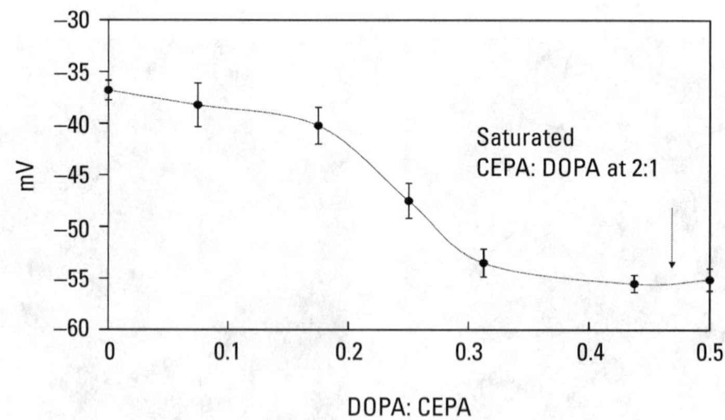

Figure 5.7 A plot of Zeta potential as a function of DOPA to CEPA ratio, which effectively measures the increasing negative charge, as CEPA is added to the surface of the particles.

5.4.2 Internal Structure of the Calcium Phosphate Shell

The overall shape of the nanoreactor appears spherical both in TEM and AFM, and the thickness of the shells appears to be dependent on the duration of supersaturation dependent shell formation. The internal structure of porous Ca/P shells appears to have only been revealed through the results of energy dispersive scattering (EDS) obtained from a PGT eXcalibur detector attached to a Hitachi H-8100 TEM at Northwestern University's EPIC facility. The X-ray diffraction (XRD) ring patterns were obtained via TEM using the Hitachi H-8100 and demonstrates the absence of structure in the shell of nanoreactor [Figure 5.8(a)], while the presence is evident in hydroxyapatite crystals [Figure 5.8(b)].

5.4.3 Concentrations of Reactants in Nanoreactors

Absorption spectra of nanoreactors filled with luminol, hematin, and fluorescein in PBS at pH 9.5, and the empty nanoshell suspensions are shown in Figure 5.9. Assuming the molar absorptivity of luminol to be $\varepsilon_{350} =$ 6,741, hematin $\varepsilon_{400} = 54{,}933$, and fluorescein $\varepsilon_{490} = 67{,}188$ [1, 13], their estimated molar concentrations are 5.2×10^{-5} M, 1.8×10^{-6} M, and 3.4×10^{-5} M, respectively. The number of nanoreactors in suspension is approximately in the nanomole range, and for the nanoreactor with 120 nm in

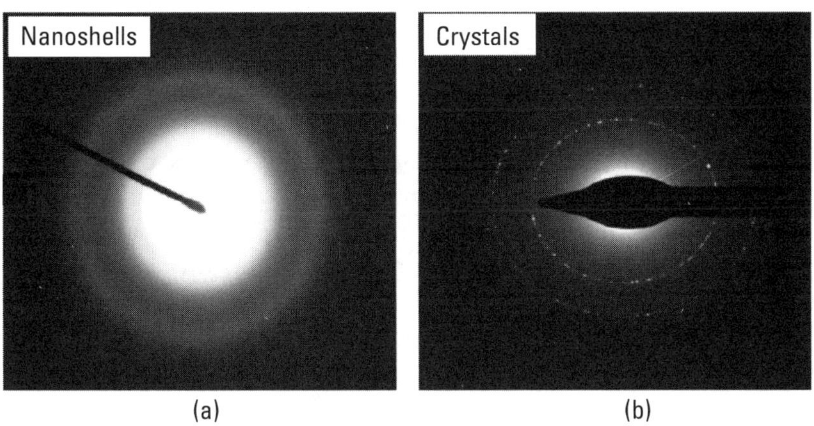

Figure 5.8 (a) Diffraction patterns obtained via TEM for liposome-templated nanoshells, and (b) commercial hydroxyapatite crystals. Note the amorphous structure of (a), whereas the presence of well-defined crystalline ring structure in (b).

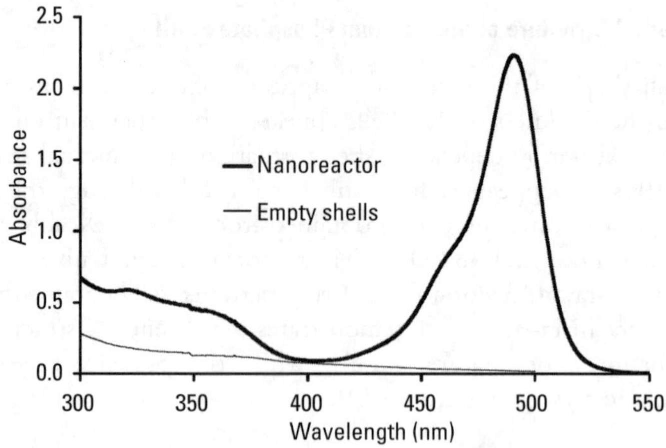

Figure 5.9 Absorption spectra of a nanoreactor suspension (heavy line) and a water-filled nanoreactor shell suspension without reagents (thin line) at pH 9.5 used to estimate concentrations of the reagents.

diameter, the concentrations of luminol and fluorescein inside nanoreactors are approximately 0.5 and 0.3 mM for luminol and fluorescein, respectively, for this preparation. Being able to maintain relatively high concentration of reactants inside the nanoreactors is an important feature to assure high efficiency of RET.

5.5 Detection of ROS with the Nanoreactor

The main objectives of this section are to demonstrate that encapsulation of the reactants for chemiluminescence and RET improves the efficiency of RET for effective detection of ROS in biological environment, as well as the overall ROS detection sensitivity. Relatively high concentrations of ROS, mostly H_2O_2, were used to compare the results in solution and in nanoreactors.

5.5.1 Stopped Flow Analyses of Luminescence

Time-resolved measurements of luminescence were performed using an Aminco-Morrow stopped-flow device (American Instrument, Silver Spring, Maryland). Appropriate concentrations of nanoshell suspension and a 5-mM H_2O_2 solution in PBS at pH 9.5 were each placed in one of the two pneumatically driven mixing syringes of the apparatus. The resulting light emission was detected using a Hamamatsu R928 photomultiplier (PMT) tube (Hamamatsu) and amplified using a Keithley current amplifier (Cleveland, Ohio). The signals were digitized at a sampling rate of 5 to 100 Hz using a LabView I/O card (National Instruments, Austin, Texas), and recorded via a homemade LabView 6i program to display the relative intensity of CL over time. At least three traces were averaged to display the luminescence change over time.

5.5.2 Time-Resolved Luminescence of Luminol in Solution and Inside Nanoreactors

The kinetics of light emission within the range of wavelengths detectable by the stopped-flow apparatus from nanoreactor suspension and equivalent concentrations of luminol, hematin, and fluorescein in PBS at pH 9.5 mixed with an equal volume of 5-mM H_2O_2 in PBS at pH 9.5 are shown in Figure 5.10. The initial rise of luminescence intensity of nanoreactor is nearly as fast as that of luminescence in solution, suggesting that the rate of H_2O_2 permeation through the shell is almost equal to the rate of diffusion in solution, and the initial intensity is higher in nanoreactor over the equivalent system in solution. The time-resolved luminescence are the result of both chemiluminescence and fluorescence, and the total emission as determined by integration of the exponentially decreasing intensity over time appears to be

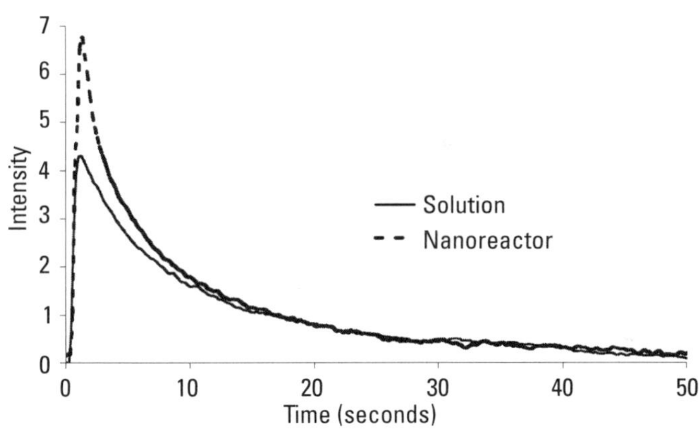

Figure 5.10 Time-resolved total luminescence from the equivalent concentration of luminol and fluorescence in solution (solid line) and in the nanoreactor (dotted line) at pH 9.5 in response to 5-mM H_2O_2 when mixed with a stopped-flow apparatus. Note the overall increase in luminescence found in the nanoreactor and the similar response time between the two samples.

larger in nanoreactor than it is for the equivalent reagents in solution. Since the total luminescence intensity is defined by the consumption of luminol, which is equal in both cases, the increased luminescence may be a reflection of overall sensitivity of the nanoreactor. This conclusion is still supported even though the relative emission intensities at the wavelengths for chemiluminescence and for fluorescence are not adjusted relative to each other.

5.5.3 Spectrophotometric Chemiluminescence and Fluorescence Analyses Show That RET Is Significantly Enhanced in Nanoreactors

Luminescence emission spectra were obtained using a SPEX SLM Model 8100 spectrofluorometer (Edison, New Jersey). Luminescence emission spectra of 1-ml nano–CRET suspension or a solution with equivalent reagents in PBS at pH 9.5 were obtained by manually mixing with 1 ml of 5 mM M H_2O_2 in the same PBS in a quartz cuvette. The scanning of emission wavelengths was started immediately after mixing at a stepping rate of 12 nm/second over the range of 350 to 650 nm with the excitation lamp turned off (Figure 5.11). Luminescence emission spectra of nanoreactor suspension (heavy line) and a solution of CRET reactants (thin line) are shown over the

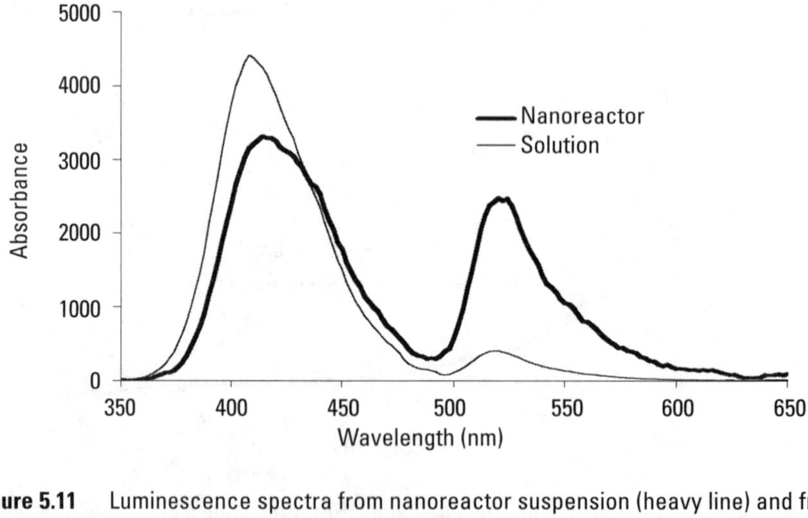

Figure 5.11 Luminescence spectra from nanoreactor suspension (heavy line) and from solution (thin line) both having the equivalent concentrations of reactants after reacting with 0.1 M H_2O_2. In the solution, most of the energy appears as chemiluminescence, but in the nanoreactor suspension, a significant amount of energy has been shifted to fluorescence, suggesting improved efficiency of RET.

range of 350 to 650 nm, and the spectral intensities have been corrected for the exponential time-dependent decrease in the emission during the time of intensity measurement from 350 to 650 nm. Both luminol chemiluminescence at 410 nm and fluorescein fluorescence at 530 nm are evident in each spectra, since some fraction of luminol show chemiluminescence and another fraction of excited-state 3AP* engage in resonance energy transfer with fluorescein and emits light as fluorescence from fluorescein. The fluorescence intensity is significantly greater in nanoeactor than in solution, while the chemiluminescence intensity is decreased, showing that resonance energy transfer is enhanced in nanoreactor.

The enhanced fluorescence emission can be justified by considering the average distance between reacting molecules inside the nanoshells. For successful energy exchange the distance of the donor and acceptor in the RET couple must be at or near their respective Forster radii, approximately 1 to 10 nm. If we consider luminol, hematin, and fluorescein to be equivalent, the nanoshell contains 8.8×10^{-5} M encapsulated molecules. The approximate concentration of nanoshells is ~1 nM of particles, which means that 88,000 molecules are present in each particle. Approximating the average volume of a single 150-nm nanoshell of volume 1.413×10^{-20} m^3 as a cube of dimension 2.42×10^{-7} m^3 on each side, a cubic array of 44 equally spaced molecules would be distributed throughout this cube. This would mean an average separation between each molecule of about 6 nm, well within the 1 to 10 nm Forster resonance energy transfer radius of most molecules.

5.5.4 The RET Takes Place Inside Nanoreactors

In order to demonstrate that the fluorescence of fluorescein at 530 nm is the result of resonance energy transfer between 3AP* and fluorescein both inside the nanoshells, the emission spectra of nanoshells containing only luminol and hematin reacted with H_2O_2 in PBS at pH 9.5 is compared with the same nanoshells suspended in a solution of fluorescein in PBS at pH 9.5 then reacted with H_2O_2 (Figure 5.12). The molar ratio between luminol inside the nanoshells and fluorescein in the surrounding solution is kept at 1:1 and the spectra are compared after correcting for dilution. Both samples show chemiluminescence at 425 nm, but fluorescein emission at 530 nm is absent, except very small luminescence, which could have been triggered with a contamination of chemiluminescence system on the surface of the shells, demonstrating that resonance energy transfer across the shell of the nanoshells is absent.

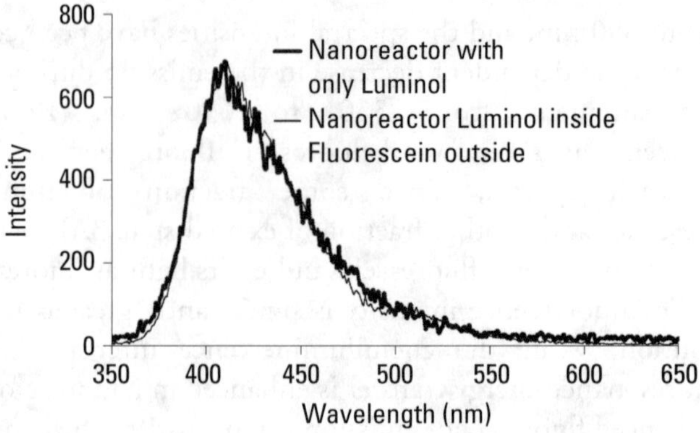

Figure 5.12 Emission spectrum from nanoshell suspensions filled with luminol and hematin (heavy line) and the same nanoshell suspensions bathed in fluorescein solution (thin line). Both reacted with 0.1 M H_2O_2. The overall ratio of encapsulated luminol to free fluorescein is about 1:1 in both cases. The fact there is little evidence of fluorescence through RET from fluorescein suggests that practically no RET takes place across the shell walls.

5.6 Reactive Oxygen Species (ROS) and Diseases

5.6.1 Significance of ROS in Human Bodies

The balanced production and consumption of each ROS is key to good health. For example, low concentrations of H_2O_2 are used in the body to regulate cell development, cell proliferation, cell death, and signal transduction [14] by directly affecting other redox systems like NADPH/NADP, ascorbic acid/dehydroascorbate, glutathione/oxidized glutathione, thioredoxin/oxidized thioredoxin, and protein carbonylation, and disulfide bond formation. Higher concentrations not adequately regulated by catalase, glutathione peroxidase, and other chemical reactions can be toxic. During infection and disease, increases in overall levels of ROS are stimulated by a variety of chemoattractant peptides, chemokines, foreign substances, and bacteria which induce neutrophils, macrophages, and even nonphagocytic cells to increase production of $O_2^{\bullet-}$. ROS levels in the body are now known to be associated with diseases, such as circulatory infection by *H. pylori* [15] (7) general bacterial infections [16–18], hypertension [19–24, 33], cancer [25–28], aging [29–31], and diabetes [34]. Adverse symptoms can arise directly from the cytotoxic properties of ROS, or by their indirect impact on metabolic regulation. For example, in conditions like hypertension, ROS production by vascular cells in various organs is increased, which lowers nitric oxide ($NO^{\bullet}$) bioactivity, interferes with nitric oxide signaling, and leads to blood pressure disregulation and eventual coronary heart disease.

Several species of ROS are found in tissues and bloodstream, including highly reactive super oxide ($O_2^{\bullet-}$), and the relatively more stable secondary oxidants derived from it via enzymatic or chemical transformations: hydrogen peroxide (H_2O_2), hydroxyl radical ($OH^{\bullet}$), hypochlorous acid (HOCl), and peroxynitrate (ONOO). In neutrophils, macrophages, and nonphagocytic healthy mammalian cells superoxide, $O_2^{\bullet-}$ is formed through the enzymatic action of NADPH oxidase, uncoupled nitric oxide synthase, xanthine oxidase, cytochrome P450 mixed–function oxidases (liver), and mitochondrial respiration [34, 35]. In addition to reacting with and damaging complex cellular molecules such as fats, proteins, or DNA, $O_2^{\bullet-}$ quickly forms H_2O_2 through the action of super oxide dismutase: $2\ O_2^{\bullet-} + 2H^+ \rightarrow H_2O_2 + O_2$, $OH^{\bullet}$ by the reaction of superoxide with H_2O_2 and Fe^{3+}/Fe^{2+}-mediated Fenton reaction $O_2^{\bullet-} + H_2O_2 \rightarrow O_2 + OH + OH^-$, HOCl by the reaction of myeloperoxidase with H_2O_2 in the presence of chlorine anions, and ONOO by the reaction with $NO^{\bullet}$. Nitric oxide ($NO^{\bullet}$) is not produced directly from $O_2^{\bullet-}$, but is produced by nitric oxide synthase in the presence of molecular oxygen, during the conversion of the amino acid L-arginine to nitric oxide and L-citrulline. In general, ROS other than H_2O_2 have short lifetimes making an estimate of their true production rates in blood using current technology challenging.

5.6.2 Conventional Methods of ROS Detection Are Cumbersome and Often Error Ridden Due to the Influence of Compounds Found in the Body

Enzymatic assays for the detection of ROS using horseradish peroxidase or catalase depend on the oxidation of a detector compound [22], but natural biological substances may lead to underestimation of ROS levels by as much as 100% [34]. Given the time required for removal of biological components before detection [35] and conditions where glutathione peroxidase-reductase significantly contributes to metabolism of H_2O_2, calculated concentrations will be underestimated [36].

Spectroscopic measurements, particularly chemiluminescence (CL) are preferred methods for oxidant detection over enzymatic and electrochemical methods in solutions due to their sensitivity, minimal perturbation of the sample, and potential for remote detection. CL is an emission of light without excitation energy from specific classes of molecules, chemically activated at the level of picomolar oxidant concentrations. CL using luminol-based reagents and lumigen-enhanced methods [37, 38] are being used to provide important information about the oxidative state of cells and tissues [39–41], loss of neutrophil activity by *H.pylori* in blood [15], during the respiratory burst of macrophages and neutrophils [42], and to diagnose vascular dysfunctions [43]. A wide variety of derivatives and conjugates of luminol [44–49] or isoluminol [50, 51], which retain their CL capability can also be synthesized. These reagents have been used to measure the concentration of a wide range of oxidants in gases, water [52], oxidant-generating enzyme solutions [53], and cells [42, 54, 55], making CL a routine method for HPLC analyte detection [56, 57].

The fact that CL reagents interact strongly with other molecules in their surroundings can be both an advantage and a disadvantage. For instance, CL emission can be competitively inhibited by biomolecules with sulfhydrl and thioether groups, metalloenzymes, horse-radish peroxidase, hematin, sugars, and even bicarbonate. Furthermore, luminescence in the visible range may be obscured by myoglobin in the muscle and hemoglobin in erythrocytes. In addition to ROS, other oxidants such as ferricyanide, hypochlorite, and persulfate can stimulate CL emission. L-012, a derivative of luminol, was found to be bright enough to be detected in highly vascularized tissues and serum-based cell cultures, but its emission is affected other chemicals including uric acid, deferoxamine, DMSO, mannitol, and sodium azide, and is light sensitive, so measurements are made in the dark [58]. Interactions between CL reagents and other molecules can be taken advantage of to shift the emissive wavelength of detection to a more convenient one, by adding a resonant energy acceptor, or sensitizer molecule like fluoroscein, rhodamine, bilurubin, riboflavin, and many others, which can increase CL output as well as shift its energy to a region of the light emission spectrum that is more amenable for detection. Microheterogenous suspensions of liposomes and micelles can be used to favorably orient and fix the positions of the CL and RET species to enhance the efficiency of their interactions [59]. However, in complex and variable biological fluids, liposomes and micelles are difficult to maintain and other unpredictable effects on CL output and interactions with sensitizers can result.

5.7 Conclusions

Spectrophotometric procedures are one of the most sensitive analytical methods requiring relatively simple operations and widely used in many areas of disciplines, but their direct applications in the biological system may be seriously affected by the materials found in them. Discussed in this chapter is a use of reordered energy transfer inside recently developed nano-enclosures to residue environment while enhancing the detection efficiency.

References

[1] Neufeld, H. A., Conklin, C. J., and Towner, R. D., "Chemiluminescence of Luminol in the Presence of Hematin Compounds," *Analytical Biochemistry*, Vol. 12, 1965, pp. 303–309.

[2] Olsson, T., Ewetz, L., and Thore, A., "Catalytic Action and Destruction of Protohematin During Peroxide Dependent Luminol Chemiluminescence," *Photochem. Photobiol.,* Vol. 38(2), 1983, pp. 223–229.

[3] Xu, G., and Dong, S., "Chemiluminescent Determination of Luminol and Hydrogen Peroxide Using Hematin Immobilized in the Bulk of a Carbon Paste Electrode," *Electroanalysis,* Vol. 11(16), 1999, pp. 1180–1184.

[4] Förster, T, "Zwischenmolekulare Energiewanderung und Fluoreszenz," *Ann. Phys.* Vol. 2, 1948, pp. 55–75.

[5] Dexter, A., "Theory of Sensitized Luminescence in Solids," *J. Chem. Phys.,* Vol. 21, 1953, pp. 836–850.

[6] Srinivas, G. and Bagchi, B., "Effect of Orientational Motion," *J. Phys. Chem. B,* Vol. 105, 2001, pp. 9370–9374.

[7] Lakowicz, J. R., Szmacinski, H., Cryczynski, I., Wiczk, W., and Johnson, M., "Influence of Diffusion on Excitation Energy Transfer in Solutions by Gigahertz Harmonic Content Frequency-Domain Fluorometry," *J. Phys. Chem.* Vol. 94, 1990, pp. 8413–8416.

[8] Yun, C., Javier, A., Jennings, T., Fisher, M., Hira, S., Peterson, S., Hopkins, B., Reich, N. O., and Strouse, G. F., "Nanometal Surface Energy Transfer in Optical Rulers: Breaking the Fret Barrier," *J. Amer. Chem. Soc.,* Vol. 127, 2005, pp. 3115–3119.

[9] Lakowicz, J. R., Piszczek, G., and Kang, J. S., "On the Possibility of Long-Wavelength Long-Lifetime High-Quantum-Yield Luminophores," *Anal Biochem,* Vol. 288, 2001, pp. 62–75.

[10] Eggeling, C., Kask, P., Winkler, D., and Jäger, S., "Rapid Analysis of Förster Resonance Energy Transfer by Two-Color Global Fluorescence Correlation Spectroscopy: Trypsin Proteinase Reaction," *Biophys. J. BioFAST,* April 22, 2005.

[11] Schmidt, H., et al., "Calcium Phosphate Based Nanoshell for Use in Biomedical Applications," Dissertation, University of Notre Dame, 2006.

[12] Schmidt, H., Kroczynski, M., Maddox, J., and Ostafin, A. E., "Antibody Conjugated Soybean Oil-Filled Calcium Phosphate Nanoshells for Targeted Delivery of Hydrophobic Molecules," *Journal of Microencapsulation,* Vol. 23(7), 2006, pp. 769–781.

[13] Dawson, R. M. C., Elliot, D. C., Elliot, W. H., and Jones, K. M., (eds.), *Data for Biochemical Research*, 3rd ed., Oxford, U.K.: Oxford Science, 1985.

[14] Fialkow, L., Wang, Y., and Downey, G., "Reactive Oxygen and Nitrogen Species as Signaling Molecules Regulating Neutrophil Function," *Free Radical Biology & Medicine*, Vol. 42(2), 2007, pp. 153–164.

[15] Mashimo, M., .Nishikawa, M., Higuchi, K., Hirose, M., Wei, Q., Haque, A., Sasaki, E., Shiba, M., Tominaga, K., Watanabe,T., Fujiwara, Y., Arakawa, T., and Masayasu, I., "Production of Reactive Oxygen Species in Peripheral Blood is Increased in Individuals with Helicobacter pylori Infection and Decreased After Its Eradication," *Helicobacter.*, Vol. 11, 2006, pp. 266–271.

[16] Akamatsu, H., Horio, T., and Hattori, K., "Increased Hydrogen Peroxide Generation by Neutrophils from Patients with Acne Inflammation," *Int. J. Dermatol.*, Vol. 42, 2003, pp. 366–369.

[17] Pericone, C. D., Overweg, K., Hermans, P. W. M., and Weiser, J. N., "Inhibitory and Bactericidal Effects of Hydrogen Peroxide Production by Streptococcus Pneumoniae on Other Inhabitants of the Upper Respiratory Tract," *Infection Immunity*, Vol. 68, 2000, pp. 3990–3997.

[18] Tomas, M. S. J., Otero, M. C., Ocana, V., and Nader-Macias, M. E., "Production of Antimicrobial Substances by Lactic Acid Bacteria. I. Determination of Hydrogen Peroxide," *Methods in Molecular Biology*, Vol. 268, 2004, pp. 337–346.

[19] Ye, S., Zhong, H., and Campese, V. M., "Oxidative Stress Mediates the Sympathetic Nerve Activity in the Phenol Renal Injury of Hypertension," *Hypertension*, Vol. 48, 2006, pp. 309–315.

[20] Tarpey, M., Wink, D., and Grisham, M., "Methods for the Detection of Reactive Metabolites of Oxygen and Nitrogen: In Vitro and In Vivo Considerations," *Am. J. Physiol. Regul. Integr. Comp. Physiol.*, Vol. 286, 2004, pp. 431–444.

[21] Tepel, M., "Oxidative Stress: Does It Play a Role in the Genesis of Essential Hypertension and Hypertension of Uraemia?" *Nephrol. Dial Transplant*, Vol. 18, 2003, pp. 439–442.

[22] Tarpey, M., and Fridovich, I., "Methods of Detection of Vascular Reactive Species Nitric Oxide, Superoxide, Hydrogen Peroxide, and Peroxynitrite," *Circulation Research*, Vol. 89, 2001, pp. 224–236.

[23] Lacy, F., Kailasam, M. T., O'Connor, D. T., Schmid-Schönbein, G. W., and Parmer, R. J., "Plasma Hydrogen Peroxide Production in Human Essential Hypertension," *Hypertension*, Vol. 36(5), 2000, pp. 878–884.

[24] Kerr, S., Brosnan, M. J., McIntyre, M., Reid, J. L., Dominczak, A. F., and Hamilton, C. A., "Superoxide Anion Production Is Increased in a Model of Genetic Hypertension: Role of the Endothelium," *Hypertension*, Vol. 33, 1999, pp. 1353–1358.

[25] Banerjee, D., Madhusoodanan, U. K., Nayak, S., and Jacob, J., "Urinary Hydrogen Peroxide: A Probable Marker of Oxidative Stress in Malignancy," *Clinica Chimica Acta.*, Vol. 334(1-2), 2003, pp. 205–209.

[26] Ran, S., Downes, A., and Thorpe, P. E., "Increased Exposure of Anionic Phospholipids on the Surface of Tumor Blood Vessels," *Cancer Research*, Vol. 62(21), 2002, pp. 6132–6140.

[27] Shmielau, J., and Finn, O. J., "Activated Granulocytes and Granulocye-Derived Hydrogen Peroxide Are the Underlying Mechanism of Suppression of T-Cell Function in Advanced Cancer Patients," *Cancer Res.*, Vol. 61, 2001, pp. 4756–4760.

[28] Kaneko, T., Yamashima, T., Tohma, Y., Nomura, M., Imajoh-Ohmi, S., Saido, T. C., Nakao, M., Saya, H., Yamamoto, H., and Yamashita, J., "Calpain-Dependent Proteolysis of Merlin Occurs by Oxidative Stress in Meningiomas: A Novel Hypothesis of Tumorigenesis," *Cancer*, Vol. 92(10), 2001, pp. 2662–2672.

[29] Kumar, D., Moore, R. M., Elkhwad, M., Silver, R. J., and Moore, J. J., "Vitamin C Exacerbates Hydrogen Peroxide Induced Apoptosis and Concomitant PGE2 Release in Amnion Epithelial and Mesenchymal Cells, and in Intact Amnion," *Placenta*, Vol. 25(6), 2004, pp. 573–579.

[30] Kukovetz, E., Bratschitsch, G., Hofer, H. P., Egger, G., and Schaur, R. J., "Influence of Age on the Release of Reactive Oxygen Species by Phagocytes as Measured by a Whole Blood Chemiluminescence Assay," *Free Radical Biol. Med.*, Vol. 22, 1997, pp. 433–438.

[31] Hagihara, H., Nomoto, A., and Mutoh, S., "Role of Inflammatory Response in Initiation of Atherosclerosis: Effects of Antiinflammatroy Drugs on Cuff-Induced Leukocyte Accumulation and Minimal Thickening of Rabbit Carotid Artery," *Atherosclerosis*, Vol. 91, 1991, pp. 107–116.

[32] Gongora, M., Qin, Z., Laude, K., Kim, H., McCann, L., Folz, J., Dikalov, S., Fukai, T., and Harrison, D., "Role of Extracellular Superoxide Dismutase in Hypertension," *Hypertension*, Vol. 48(3), 2006, pp. 473–481.

[33] San Martin, A., Du, P., Dikalova, A., Lassegue, B., Aleman, M., Gongora, M., Brown, K., Joseph G., Harrison, D., Taylor, W., Jo, H., and Griendling, K., "Reactive Oxygen Species-Selective Regulation of Aortic Inflammatory Gene Expression in Type 2 Diabetes," *American Journal of Physiology. Heart and Circulatory Physiology*, Vol. 292(5), 2007, pp. H2073–2082.

[34] Boveris, A., Martino, E., and Stoppani, A. O., "Evaluation of the Horseradish Peroxidase-Scopoletin Method for the Measurement of Hydrogen Peroxide Formation in Biological Systems," *Anal. Biochem.*, Vol. 80, 1977, pp. 145–158.

[35] Staniek, K., and Nohl, H., "H_2O_2 Detection from Intact Mitochondria as a Measure for One-Electron Reduction of Dioxygen Requires a Non-Invasive Assay System," *Biochim. Biophys. Acta.* Vol. 1413, 1999, pp. 70–80.

[36] Kinnula, V. L., Mirza, Z., Crapo, J. D., and Whorton, A. R., "Modulation of Hydrogen Peroxide Release from Vascular Endothelial Cells by Oxygen," *Am. J. Respir. Cell Mol. Biol.*, Vol. 9, 1993, pp. 603–609.

[37] Sohn, H. Y., Gloe, T., Keller, M., Schoenafinger, K., and Pohl, U., "Sensitive Superoxide Detection in Vascular Cells by the New Chemiluminescence Dye L-012," *J. Vasc. Res.,* Vol. 36(6), 1999, pp. 456–464.

[38] Daiber, A., August, M., Baldus, S., Wendt, M., Oelze, M., Sydow, K., Kleschyov, A. L., and Munzel, T., "Measurement of NAD(P)H Oxidase-Derived Superoxide with the Luminol Analogue L-012," *Free Radic. Biol. Med.*, Vol. 36(1), 2004, pp. 101–111.

[39] Kopprasch, S., Pietzsch, J., and Graessler, J., "Validation of Different Chemiluminegic Substrates for Detecting Extracellular Generation of Reactive Oxygen Species by Phagocytes and Endothelial Cells," *Luminescence*, Vol. 18, 2003, pp. 268–273.

[40] Povov, I., Volker, H., and Lewin, G., "Photochemiluminescent Detection of Antiradical Activity. V. Application in Combination with the Hydrogen Peroxide-Initiated Chemiluminescence of Blood Plasma Proteins to Evaluate Antioxidant Homeostasis in Humans," *Redox Report*, Vol. 6, 2001, pp. 43–48.

[41] Frei, B., Yamamoto, Y., Niclas, D., and Ames, B. N., "Evaluation of an Isoluminol Chemiluminescence Assay for the Detection of Hydroperoxides in Human Blood Plasma," *Anal. Biochem.*, Vol. 175(1), 1988, pp. 120–130.

[42] Imada, I., Sato, E., Miyamoto, M., Ichimori, Y., and Minamiyama, Y., "Analysis of Reactive Oxygen Species Generated by Neutriphil Using a Chemiluminescecne Proble L-012," *Anal. Biochem.*, Vol. 271, 1999, pp. 53–58.

[43] Ergul, A., Schultz Johansen, J., Strømhaug, C., Harris, A. K., Hutchinson, J., Tawfik, A., Rahimi, A., Rhim, E., Wells, B., Caldwell, R. W., and Anstadt, M. P., "Reactive Oxygen Species in Diabetes: Role of Endothelin-1," *The Journal of Pharmacology and Experimental Therapeutics*, Vol. 313, 2005, pp. 70–77.

[44] Motoyoshiya, J., Yokota, K., Fukami, T., Konno, S., Yamamoto, A., Hotta, M., Koike, R., Yoshioki, S., Nishi, Y., and Aoyama, H., "Synthesis of 4-Styrlphthalhydrazides and Their Chemiluminescence Reaction. Emitters and Chemiluminescence Efficiency Highly Depend on Electronic Nature of Styryl Groups," *Journal of Hetrocyclic Chemistry*, Vol. 42(6), 2005, pp. 1063–1068.

[45] Neumann, H., Klaus, S., Klawonn, M., Strubing, D., Hubner, S., Gordes, D., von Wangelin, A. J., Lalk, M., and Beller, M., "A New Efficient Synthesis of Substituted Luminols Using Multicomponent Reactions," *Zeitschrift fur Naturforschung B- Journal of Chemical Sciences*, Vol. 59(4), 2004, pp. 431–438.

[46] Mills, R., and Wu, G., "Synthesis and Evaluation of Novel Prodrugs of Foscarnet and Dideoxycytidine with a Universal Carrier Compound Comprising a Chemiluminescent and a Photochromic Conjugate," *Jour. Pharm.l Sci.* Vol. 93(5), 2004, pp. 1320–1326.

[47] Yoshida, H., "Development of Sensitive Liquid Chromatographic Determination Methods Using Novel Luminol Derivatives and Electrogenerated Chemiluminescence of Ruthenium Complex," *Bunseki Kagaku,* Vol. 52(3), 2003, pp. 215–216.

[48] Nishinaka, Y., Aramaki, Y., Yoshida, H., Masuya, H., Sugawara, T., and Ichimori, Y., "A New Sensitive Chemiluminescence Probe, L-012, for Measuring the Production of Superoxide Anion by Cells," *Biochem. Biophys. Res. Comm.* Vol. 193(2), 1993, pp. 554–559.

[49] Richards, S. J., and Westeman, F. S., "Alkali-Labile Luminol Derivatives as Labels for Quantitative Binding Studies With Polyionic Macromolecules," *Jour. Biolumin. Chemilumin.* Vol. 3(4), 1989, pp. 175–179.

[50] Messeri, G., Orlandini, A., and Pazzagli, M., "Luminescent Immunoassay Using Isoluminol Derivatives," *Jour. Biolumin. Chemilumin.* Vol. 4(1), 1989, pp. 154–155.

[51] Schroeder, H. R., and Yaeger, M., "Chemiluminescence Yields and Detection Limits of Some Isoluminol Derivatives in Various Oxidation Systems," *Analytical Chemistry,* Vol. 50(8), 1978, pp. 1114–1120.

[52] Yamashiro, N., Uchida, S., Satoh, Y., Morishima, Y., Yokoyoma, H., Satoh, T., Sugama, J., and Yamada, R., "Determination of Hydrogen Peroxide in Water by Chemiluminescence Detection (I) Flow Injection Type Hydrogen Peroxide Detection System," *J. Nuclear Sci. Technol.* Vol. 41, 2004, pp. 890–897.

[53] Seitz, W. R., and Neary, M. P., "Chemiluminescence and Bioluminescence Analysis," in *Contemporary Topics in Analytical and Clinical Chemistry,* D. M. Hercules, (ed.), New York: Plenum Press, 1977, pp. 49–127.

[54] Kopprasch, S., Pietzsch, J., and Graessler, J., "Validation of Different Chemiluminegic Substrates for Detecting Extracellular Generation of Reactive Oxygen Species by Phagocytes and Endothelial Cells," *Luminescence,* Vol.18, 2003, pp. 268–273.

[55] Pick, E., and Keisari, Y., "A Simple Colorimetric Method for the Measurement of Hydrogen Peroxide Produced by Cells in Culture," *J. Immunol. Methods,* Vol. 38, 1980, pp. 161–170.

[56] Yamamoto Y., and Ames, B. N., "Detection of Lipid Hydroperoxides and Hydrogen Peroxide at Picomole Levels by an HPLC and Isoluminol Chemiluminescence Assay," *Res. Radic. Biol. Med.,* Vol. 3, 1987, pp. 359–361.

[57] Yamaguchi, M., Yoshida, H., and Nohta, H., "Luminol-Type Chemiluminescence Derivatilzation Reagents for Liquid Chromatography and Capillary Electrophoresis," *J. Chromato. A,* Vol. 950, 2002, pp. 1–19.

[58] Ergul, A., Schultz Johansen, J., Strømhaug, C., Harris, A. K., Hutchinson, J., Tawfik, A., Rahimi, A., Rhim, E., Wells, B., Caldwell, R., W., and Anstadt, M. P., "Reactive Oxygen Species in Diabetes: Role of Endothelin-1," *Journal of Pharmacology and Experimental Therapeutics.,* Vol. 313, 2005, pp. 70–77.

[59] Hadijianstis, J., and Nikokavouras J., "Luminol Chemiluminescence in Micellar Media II: Energy Transfer to Fluorescein," *J. Photochem. Photobiol. A: Chem.*, Vol. 69, 1993, pp. 337–343.

6

Surface Nanoreactors for Efficient Catalysis of Hydrolytic Reactions

I. M. Okhapkin, V. I. Lozinsky, V. V. Vasilevskaya, and A. R. Khokhlov

In this chapter the notion of a surface nanoreactor is introduced and the main principles of its performance are described. A surface nanoreactor is a boundary between two immiscible phases of a nanoscale thickness that can accumulate reacting species and promote the increase of reaction rate. Two types of nanoreactors are discussed: interfaces of emulsion droplets, and polymer interfaces formed in globules and fine polymer aggregates. Hydrolytic reactions were found to significantly speed up in the nanoreactors.

6.1 Introduction

Accurate control of the chemical reaction rate is of significant fundamental and practical interest. This requires versatile and precise means that can provide both reaction acceleration, and decrease in reaction rate in case of overheating. For many decades, catalysts, both homogeneous and heterogeneous, are the principal components of most reactions. Catalysts are mostly used to speed-up the rate of reactions that do not practically run in catalyst-free conditions due to very high kinetic barriers. An alternative way to increase the reaction rate is to accumulate the reagents in a set of microreactors distributed uniformly in the reaction medium. To concentrate reagents in a limited volume, various heterogeneous systems can be used [1], such as polymer gels [2–4], dendrimers [5], micellar solutions, and microemulsions [1, 6–12].

Chemical reactions can be accelerated by two orders of magnitude in micellar solutions [1, 6–8]. Such an effect is caused by the concentrating of reactants within the micelles; it depends on the volume fraction occupied by the micelles in the solutions [6–8]. Micellar solutions can be used for accelerating reactions provided the reacting species have high affinity to micellar phase, so that absorption of the reactants can proceed. In most cases such absorption is favored both by hydrophobic and electrostatic interactions and normally micelles of ionic surfactants are most efficient. Micellar systems can promote the reactions in which substrate and catalyst are capable to be concentrated in micelles. An alternative scheme for concentrating reactive agents and reaction-efficiency control involves the use of small polymer particles that accumulate hydrophobic substrates [13–15]. In order to provide conversion of the substrate, these particles should have a hydrophobic interior and pendant catalytic groups therein. This type of conversion might be quite similar to the processes that originate from biological systems. Indeed, this in particular relates to enzyme-substrate binding processes. Several enzymes like acetylcholine esterase and pepsin contain hydrophobic groups at their active sites, which efficiently contribute to the binding of the substrate, while special functional groups promote the catalysis.

A quite prospective tool to increase the rate of catalytic reaction may be realized via interfacial adsorption where surface-active molecules tend to be concentrated at the boundaries of immiscible phases. If the adsorbing molecules can react with each other (say conversion of a substrate that can be catalyzed in homogeneous conditions), the interfacial layer becomes a reactor of nanoscale thickness, in which chemical processes speed up. In our recent papers, this layer was referred to as a "surface nanoreactor" [16–20]. The concept of surface nanoreactors is relevant for biological and medical applications. Living organisms accumulate a great number of interfaces, and their reactions might play a significant role in the metabolism. In a typical example of interfacial reactions, it is appropriate to mention pheromone identification by the antennas of insects where the latter act as a kind of surface nanoreactor that concentrates pheromones before they activate the response of the receptors. One of the major active factors of such bioanoreactors is the surface diffusion of captured pheremones to the receptors. The analogous process might be typical for receptor-hormone binding and response mechanisms, which need separate verification and consideration. Nevertheless, when a pheromone or hormone binds to a surface or interface, its concentration at the interface is markedly higher than in the three-dimensional media, which might be a significant factor in the response that follows.

Thus the phenomenon of surface nanoreactors has some trends similar to those which can be located in biosystems and may be taken into account in studies where biological interfaces are the objects.

As far as the basic model for surface nanoreactors is concerned, emulsions prove to be the simplest. In emulsions the interface of a droplet can act as a medium for the reaction, which can then be accelerated. Reactant concentrations may be several orders of magnitude higher at such interfaces than in bulk phases. Both theoretically [16, 18] and by means of computer simulation [21] it was shown that at a certain size of emulsion droplets (normally around several hundred nanometers) the reaction rate function exhibits a maximum. In Figure 6.1, a series of dependences of reaction rate versus droplet radius are presented for different adsorption energy at the droplet interface, ε (in kT units). The reaction rate (M) is normalized by the reaction rate in homogeneous emulsion-free solution (M^0). The droplet radius (R) is normalized by the surface layer width (d). The data was obtained by computer simulation (Monte Carlo method). The positions of the maxima were found to be sensitive to ε; the lower energy of adsorption required the lower optimum size of emulsion droplets.

Indeed, if the droplets are small, the surface layer area is large; consequently the concentration of reactants that are distributed over this surface is not very high compared to the volume concentration. The reaction rates as results are small. With an increase in the droplet size (and decrease in surface layer area, accordingly), the surface concentrations of the reactants grow. In turn, this promotes an increase of the reaction rate. At higher droplet sizes, the decrease in the reaction rate is observed because the growth of droplet size results in the diminution of the surface area. Therefore, the "surface" contribution to the reaction rate is small and the process mainly takes place in the bulk phases. It is worth mentioning here that possible dependence of the droplet size, both on surface activity or concentration of the reactants in [16–18, 21] was ignored.

It is interesting that the reaction rate does not depend linearly on the substrate concentration for the system described. Figure 6.2 displays the corresponding dependence. The catalyst surface activity is taken to be high: $\varepsilon_B = -12$. This means that the catalyst molecules are mostly located at the interface.

At low ε_A, both substrate concentration and reaction rate increase until substrate molecules saturate the surface layer. Further increase in the

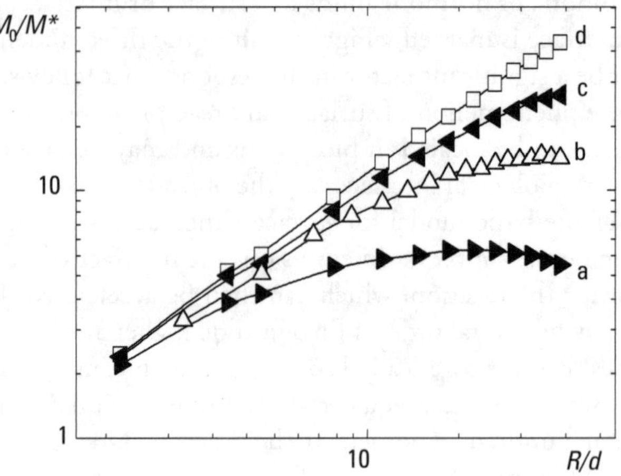

Figure 6.1 Reaction rate (M_0) dependence on droplet diameter R at equal substrate and catalyst concentrations $\varphi_A = \varphi_B = 4.8 \cdot 10^{-4}$, dispersed phase volume fraction V_{in}/V=0.1 and various surface activities of the reactants, $\varepsilon_A = \varepsilon_B = \varepsilon_S$: −4 (a); −5 (b); −6 (c); -14(d). Adapted from [20].

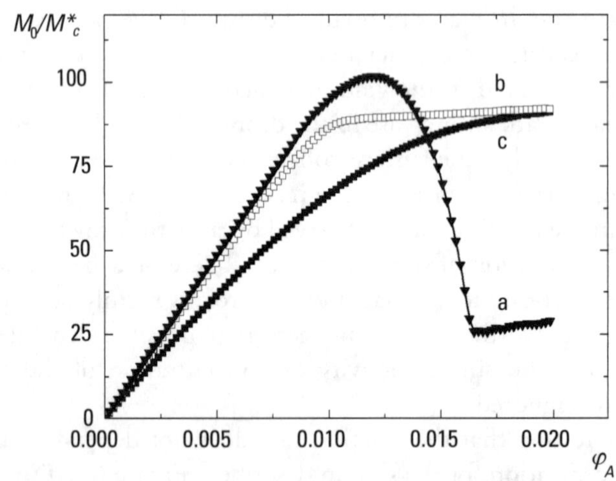

Figure 6.2 Reaction rate (M_0) dependence on substrate concentration (φ_A) at catalyst concentration φ_B=7.6·10^{-3}, catalyst surface activity ε_B=−12, dispersed phase volume fraction V_{in}/V=0.1 and various surface activities of substrate ε_A: -16 (a); -8 (b); -6(c). Adapted from [20].

substrate volume concentration leads to various nonlinear effects depending on substrate-to-catalyst surface activity ratio. If $\varepsilon_A < \varepsilon_B$, substrate molecules can displace catalyst molecules from the surface which leads to the drop of reaction rate values. Inversely, if $\varepsilon_A > \varepsilon_B$ no substitution of the catalyst by the substrate is observed. Therefore, after the surface is saturated by the substrate, the reaction rate variation stops and the dependence attains plateau (Figure 6.2) showing Michaelis-Menten–like profile. This analogy to the enzyme kinetics will be discussed in the concluding remarks.

6.1.1 Emulsion-Based Surface Nanoreactors

Computer modeling and theoretical consideration have shown a possibility to increase the rate of catalytic reactions while carrying them out in *surface nanoreactors*. Below we will dwell on the experimental realization of such possibility. First, oil-in-water emulsions [22] are considered as the source of the nanoreactors; second, nanoreactors based on fine polymer dispersions [19, 23] will now be described.

For preparation of the corresponding oil-in-water emulsion, *n*-dodecane was used as the component of the dispersed phase. Volume fraction of the hydrocarbon was 1%. The emulsion was prepared by ultrasonic homogenization at a frequency of 20 kHz. A small amount of poly (*N*-vinylpyrrolidone-*co*-acrylic acid) (25% of acrylic acid, $M_w = 96000$) was added for stabilization of the emulsion. As surface nanoreactor, the ultrathin oil-water boundary of an emulsion droplet was considered. Hydrolysis of p-nitrophenyl butyrate was chosen as a model reaction to be studied in the presence of surface nanoreactors (Scheme 1). It was possible to use the homological p-nitrophenol esters of n-carbonic acids as substrates for hydrolysis. p-Nitrophenyl butyrate was the optimal one taking into account solution and surface properties of the substrates.

Hydrolysis of p-nitrophenyl esters is easily catalyzed by imidazole moieties. To provide higher surface activity of catalyst that is required by the theory of the method, two ways can be employed. First, it is possible to synthesize a surfactant containing a long hydrophobic tail and an imidazole ring. Second, imidazole moieties can be introduced into a water-soluble polymeric surfactant. In our work, the second variant was chosen due to the special properties of polymeric surfactant to be adsorbed permanently to hydrophobic surfaces. Indeed, the adsorption of surfactants is a dynamic process: normally there is a mass exchange between surface layer and volume phase. Polymer surfactants undergo desorption and readsorption at much smaller rate than low-molecular surfactants which can diminish possible diffusional limitations arising from mass-exchange processes. To hydrolyze p-nitrophenyl butyrate a copolymer of N-vinylimidazole and N-vinylcaprolactam (PVCL-Vim, Scheme 2) (20% of N-vinylimidazole (NMR), $M_w = 15000$ (SLS)) was prepared. The preparation procedure is described elsewhere [19]. N-vinylcaprolactam groups confer surface activity to the copolymer, whereas N-vinylimidazole provide catalytic properties. Futhermore, copolymers of poly(N-vinylcaprolactam) are widely known for their specific thermoresponsive properties [24–27]. These copolymers are water soluble at low temperatures. However they exhibit a lower critical solution temperature (LCST) in water and undergo a coil-to-globule transition accompanied by aggregation at higher temperature. The resulting polymer particles possess developed surface areas which render them prospects to be used as surface nanoreactors without any supply of additional components such as emulsion droplets (this possibility is considered below).

To monitor the reaction progress, a spectrophotometric method was used since the hydrolysis of p-nitrophenyl butyrate results in the release of colored p-nitrophenolate-anion absorbing at 402 nm. The reaction was carried out in phosphate buffer medium at pH 7.4, 25°C. Phosphate concentration was 0.5 mM. Catalyst concentration was 0.05 mg/ml. After injecting the substrate into the reaction medium, the solutions were incubated at 25°C for the fixed-time period. After that period, the reaction medium was promptly filtered using Millex SLGV R33 filters before the measurement since it was impossible to register the optical density of the product in turbid emulsion. The reaction rate was calculated according to:

$$V = dD/dt \cdot \varepsilon^{-1} \cdot F^{-1} \tag{6.1}$$

where dD/dt is the initial slope of optical density variation with time, *å* is the extinction coefficient of p-nitrophenolate-anion (18 500 $L \cdot mol^{-1} \cdot cm^{-1}$)), and F is the deprotonation degree of p-nitrophenol at pH 7.4 ($F = 0.6$, as found by spectrophotometry).

The reaction intensification in emulsion can only be observed if it partially occurs in the nanoreactors formed by surface layers of the droplets. This is possible if both substrate and catalyst are surface-active and they concentrate at the surface of emulsion droplets. This is the case for both p-nitrophenyl butyrate and PVCL-Vim. If either catalyst or substrate is surface-inactive, concentrating effect of the second component should be inefficient and no intensification of the reaction should be observed.

It is well-known that p-nitrophenyl esters are subjected to spontaneous hydrolysis in aqueous solutions. This process is catalyzed by hydroxyl ions and by any other base. As the hydrolysis of p-nitrophenyl butyrate was studied in a phosphate buffer, two catalysts are possible for the reaction: hydroxyl ions and PVCL-Vim. The first one is surface-inactive, the second one is surface active. Figure 6.3 displays four dependences of p-nitrophenyl butyrate hydrolysis in various conditions. The reaction was studied both in emulsion and in homogeneous solution. In the presence of PVCL-Vim, the reaction was catalyzed both by the polymer and by hydroxyl ions. When no PVCL-Vim was introduced in the system, the reaction was catalyzed by the ions only. The data of Figure 6.3 indicate that in the presence of PVCL-Vim the reaction rate is several times higher in emulsion than in homogeneous solution. However, if spontaneous hydrolysis is considered, there is no intensification of the reaction in emulsion media compared to the case of homogeneous solution. On the contrary, a small rate drop is observed. Thus, the reaction is sped up in emulsion only in the presence of the surface-active catalyst; such an effect is not observed in spontaneous hydrolysis stimulated by surface-inactive catalysts. That is, the reactions in emulsion can be sped up only if the reacting species are surface active (that is when they can be adsorbed on interfaces). This fact is a serious support to the theoretically

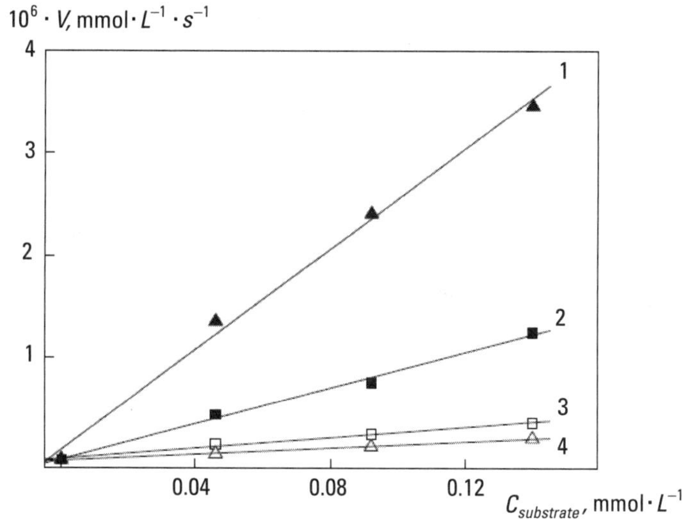

Figure 6.3 Hydrolysis of p-nitrophenyl butyrate in presence of PVCL-Vim in emulsion (1) or in homogeneous solution (2) and spontaneous hydrolysis of p-nitrophenyl butyrate in emulsion (4) and in a homogeneous solution (3). (*After:* [21].)

proposed hypothesis that disperse systems can be a medium for reaction intensification when local concentration of the reactants is increased in surface nanolayers of the dispersed particles.

As far as the drop of the spontaneous hydrolysis rate is concerned (Figure 6.3, lines 3, 4), it is possibly connected with the fact that the part of the surface-active substrate might be eliminated from the reaction media to droplet surfaces. At these surfaces, the reaction does not proceed as they are depleted of the surface-inactive catalyst.

It is worth mentioning that the contribution of spontaneous hydrolysis to the overall reaction rate is not negligible in the presence of PVCL-Vim. The data of Table 6.1 illustrate this statement. Pseudo-first-order rate constant of spontaneous hydrolysis is about 30% of that in the presence of PVCL-Vim (the processes catalyzed by basic ions and PVCL-Vim are taken to be independent and their rates are taken to be additive). Therefore, if one subtracts the spontaneous hydrolysis rate constants from the overall ones, he/she will derive the values for PVCL-Vim only. These values are presented in Table 6.1. The ratio of this values ($P = 3.7$) gives a factor of reaction rate increase in emulsion for the situation when both the substrate and catalyst can be adsorbed at phase boundaries.

Table 6.1

Kinetic Parameters of p-nitrophenyl Butyrate Hydrolysis Under Various Conditions

Process	**Reaction rate constants of pseudo-first-order, 105·k, s-1**		
	In emulsion	**In homogeneous solution**	**Factor of reaction rate increase in emulsion for PVCL-Vim**
Spontaneous hydrolysis	0.15 (k_1)	0.26 (k_2)	
Hydrolysis in the presence of PVCL-Vim (PVCL-Vim plus spontaneous hydrolysis)	2.4 (k_3)	0.87 (k_4)	P = 3.7 (P = k_6/k_4)
Contribution of PVCL-Vim only	2.25 ($k_5 = k_3 - k_1$)	0.61 ($k_6 = k_4 - k_2$)	

Source: [21].

6.1.2 Polymer-Based Surface Nanoreactors (Case of Polymer Aggregates)

The considerations that were performed for the emulsion-based surface nanoreactors are valid as well for polymer systems having developed boundaries, namely, for the dispersions of polymer associates or globules. Low-molecular weight substrates that exhibit surfactant properties can adsorb at the surface areas formed between polymer and solvent. Such surface areas are typical for polymer solutions in poor solvent conditions. Polymers undergo intra- and intermolecular aggregation in poor solvent: polymer molecules fold into compact globules which normally show a tendency to association. Furthermore, if the polymer is composed of hydrophobic and amphiphilic (or hydrophilic) monomer units, in aqueous solution the latter tend to occupy surfaces of macromolecular globules and associates. In case the hydrophilic monomer units catalyze chemical transformation of the adsorbing substrate, the polymer boundary may be considered as the surface nanoreactor with chemically bound catalytic groups.

Thermoresponsive poly(N-vinylcaprolactam-co-N-vinylimidazole) (PVCL-Vim) was found to be suitable as surface-active catalyst in emulsion-based surface nanoreactors. However, as it was stated above, the copolymer can phase separate itself upon temperature increase in aqueous solutions. This results in the formation of fine aggregates with developed surfaces. Imidazole moieties of the copolymer are much more hydrophilic than N-vinylcaprolactam units [28], and should be preferentially located at the surfaces of the aggregates. Thus, such aggregates were prospective to be tested as surface nanoreactors.

We employed poly(N-vinylcaprolactam-co-N-vinylimidazole) as catalyst for hydrolysis of p-nitrophenyl acetate (Scheme 1) in both dissolved and aggregated states. Besides poly(N-vinylcaprolactam-co-N-vinylimidazole), poly(N-isopropylacrylamide-co-N-vinylimidazole) (Scheme 2) was also investigated. This copolymer exhibits thermoresponsive properties in water solutions as well. It was expected that the copolymers should be more efficient catalysts when aggregated.

The polymer synthesis has been described elsewhere [19]. Initial monomer solutions in absolute ethanol (30 vol% of monomers) containing 2,2'-azoisobisbutyronitrile as an initiator were incubated in argon atmosphere at 50°C for 48 hours. Then the reaction solution was diluted with ethanol and poured into diethyl ether. White precipitate was collected. It was dissolved in deionized water, dialyzed, and freeze-dried. The characteristics of the polymers obtained are presented in Table 6.2. Kinetic measurements

Table 6.2
Characteristics of the Polymers Tested in Hydrolysis of p-nitrophenyl Acetate

	M_W	$T_{transition}$	N-vinylimidazole content in % Mol, titration	N-vinylimidazole content in % Mol, NMR
Poly(N-vinylimidazole) (PVim)	46,000	—	—	—
PNIPA-Vim-11[a]	48,000	35	—	11
PNIPA-Vim-13[a]	79,000	35	—	13
PVCL-Vim-20[a]	32,000	40	19	20
PVCL-Vim-29[a]	41,000	42	28	30

[a]The polymers are numbered according to the N-vinylimidazole content; PNIPA-Vim means poly(N-isopropylacrylamide-co-N-vinylimidazole). *Source:* [18].

were performed according to a procedure which was specially developed for the studied polymer solutions [18].

As the aggregates of the copolymers were formed upon raising the temperature of solution, the effects of aggregation and temperature on the catalytic properties of the copolymers overlapped. Therefore, the correlation of the catalytic properties with the aggregation was investigated using the reaction rate versus temperature dependencies, which normally give a linear plot on semilogarithmic (Arrhenius) coordinates. Possible influence of the aggregation should result in a deviation from the linear law.

Figure 6.4 shows the Arrhenius dependencies for the four copolymer catalysts, 1-methylimidazole and poly(N-vinylimidazole) (Scheme 2) as reference samples, at identical concentrations of imidazole groups. For 1-methylimidazole and poly(N-vinylimidazole), the dependencies were quite linear, showing that those catalysts followed the Arrhenius-type behavior. For copolymer catalysts, the rate-temperature dependencies were not linear in Arrhenius coordinates. In the temperature range 35°C to 45°C, the growth law was faster than the linear one. When the temperature was raised further, the opposite effect was observed, namely, the reaction rate slowed down.

The relation between the catalytic properties and aggregation of the copolymers was illustrated using the dynamic light-scattering method. Hydrodynamic diameter distributions obtained from light-scattering data showed that at temperatures below 35°C the copolymers existed in the state of coils, while aggregates were formed upon heating above that temperature

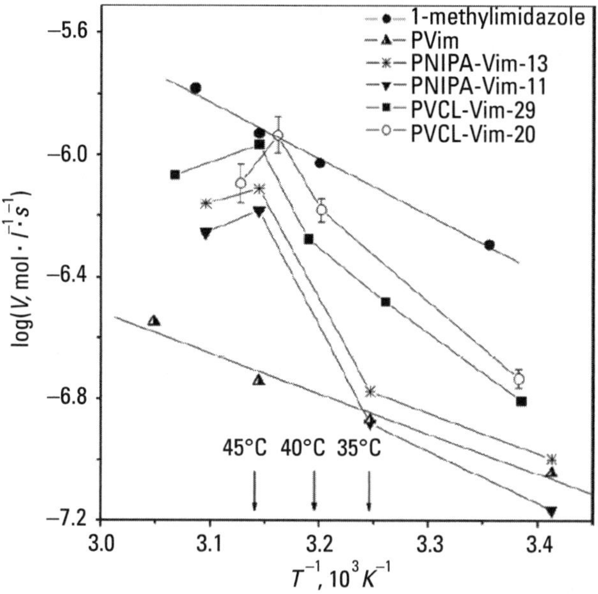

Figure 6.4 Reaction rate of hydrolysis of p-nitrophenyl acetate as a function of inverse temeperature. Thermosensitive imidazole containing copolymers (PVCL-Vim, PNIPA-Vim), 1-methylimidazole, and poly(N-vinylimidazole) act as catalysts. (*After:* [18].)

(Figure 6.5). At low temperatures the average diameter of the polymer particles did not exceed 20 nm. Upon heating, new peaks at 100 to 200 nm emerged accounting for the polymer aggregates. For all the copolymers studied, the temperature intervals of aggregation preceded those of rapid growth of the reaction rate in the region 35°C to 45°C. Thus, the observed intensification of the reaction was found to be closely connected with the aggregation phenomenon in solutions of the thermosensitive copolymers; the assumption is justified that polymer aggregates can act as the surface nanoreactors promoting the reaction.

Furthermore, a Michaelis–Menten profile of the catalyzed reaction was observed for the thermosensitive copolymers studied (Figure 6.6). In enzymatic catalysis, the catalytic act is preceded by a complex formation between catalyst and substrate. Because of the complex formation, enzymatic reactions follow Michaelis–Menten–type kinetics according to:

$$V = \frac{V_0[S]}{K_m + [S]} \tag{6.2}$$

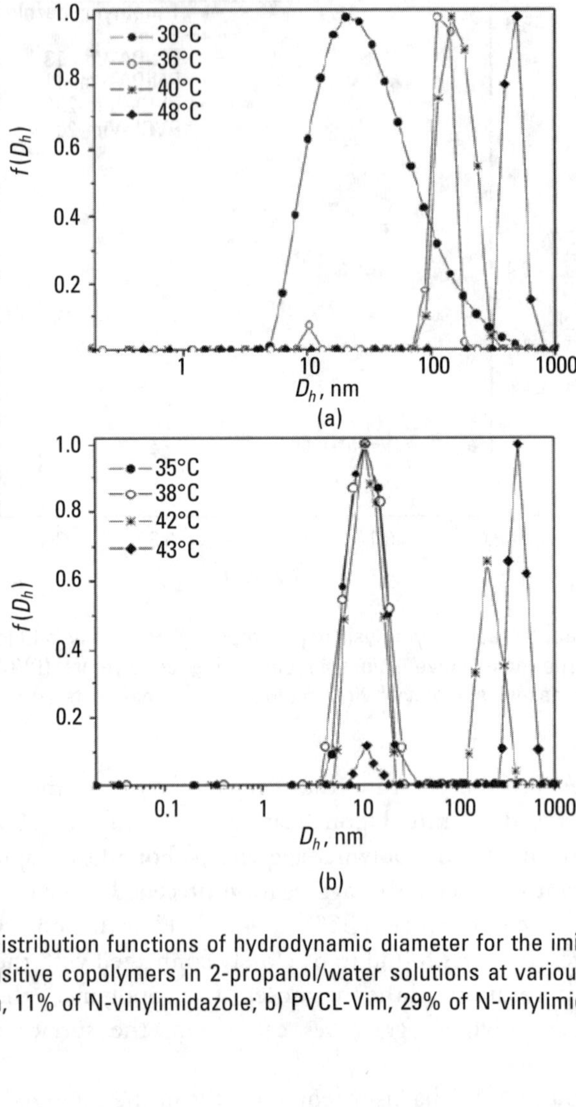

Figure 6.5 Distribution functions of hydrodynamic diameter for the imidazole-containing thermosensitive copolymers in 2-propanol/water solutions at various temperatures: (a) PNIPA-Vim, 11% of N-vinylimidazole; b) PVCL-Vim, 29% of N-vinylimidazole. Adapted from [18].

where $V_0 = k_{cat}[E_0]$, k_{cat} is the first-order-rate constant for breakdown of the substrate–catalyst complex, $[E_0]$ is the concentration of the catalyst, $[S]$ is the concentration of the substrate, and K_m is the Michaelis constant, which is the dissociation constant of the enzyme-substrate complex. For a PNIPA-Vim copolymer containing 11% of imidazole groups, a kinetics curve in V–$[S]$ coordinates was obtained which could be well fitted with (6.2) giving $V_0 = 8.6 \cdot 10^{-4}$ mmol·L^{-1}·s^{-1} and $K_m = 10.5$ mmol·L^{-1}. This

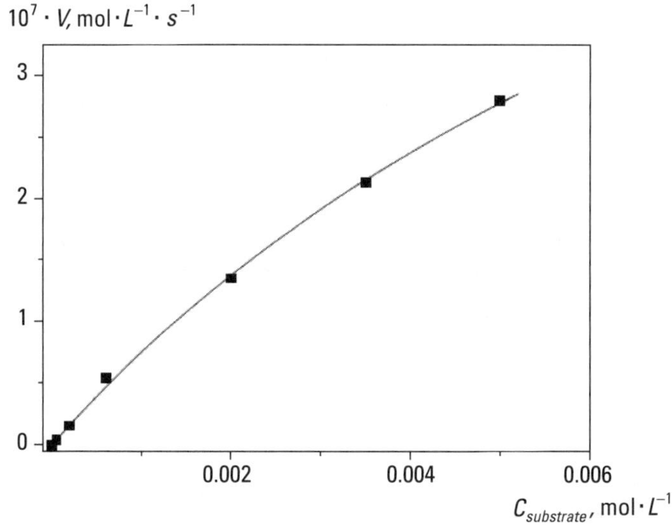

Figure 6.6 Michaelis-Menten profile for hydrolysis of p-nitrophenyl acetate in presence of PNIPA-Vim-11 at 45°C. (*After:* [18].)

additionally confirms that polymer aggregates can intensify the reaction of hydrolysis via adsorption of the substrate that form a kind of complex with them.

6.1.3 Polymer-Based Surface Nanoreactors (Case of Polymer Globules)

The copolymers described above undergo intermolecular aggregation in aqueous solution upon temperature increase; the boundaries of the aggregates formed can act as surface nanoreactors. A generous prospective of further use of such copolymers consisted in suppressing intermolecular aggregation so that only globule formation or very slight aggregation could proceed when solvent quality gets poorer. This prospective is realizable by using so-called protein-like copolymers. This concept was first introduced in [29, 30] to denote macromolecules which can assume conformation similar to that of globular proteins. Such macromolecules should be composed of two hydrophobic and hydrophilic monomer units bound in a specific sequence that enables formation of globules with hydrophobic core and hydrophilic shell. This structure prevents the globules from aggregation. The subsequent development of the concept resulted in theoretical prediction of some properties that should be characteristic for real polymers with protein-like sequences [31].

Along with the theoretical studies, several successful attempts at preparative synthesis of protein-like copolymers have been reported [32]. The syntheses were carried out using either chemical modification of polymeric precursors [33] or copolymerization [34–37] or copolycondensation [38] of monomers with distinct hydrophobicities. In particular, we prepared copolymers that had pronounced protein-like properties [34, 37, 39] by free-radical copolymerization of N-vinylcaprolactam and N-vinylimidazole in water-DMSO mixture at a temperature above the phase transition threshold of the resulting polymers (precipitation polymerization at 65°C). It was shown that the target protein-like macromolecules were only formed in a very narrow range of initial monomer ratios and that the process yielded a mixture of polymer fractions that differed in water solubility at elevated temperature. They were referred to as thermally precipitating (tp-65) and thermally nonprecipitating (ts-65) fractions. In view of the physicochemical characteristics, the latter was identified as consisting of protein-like macromolecules [37, 39]. Along with this, control samples were synthesized in the same solvent at a temperature below the phase transition point (solution polymerization). In this experiment, two polymer fractions have been obtained, too. The first one precipitated upon temperature increase (tp-25), the second one (ts-25) was not sensitive to temperature variation. The monomer composition of the four fractions was quite similar as well as the monomer feed ratios. Figure 6.7 illustrates the properties of those fractions. Conformational transitions of N-vinylcaprolactam copolymers can be monitored by high sensitivity differential scanning calorimetry [27, 39]. Normally, thermograms obtained during heating of the copolymers show a single broad and asymmetric peak of heat capacity, observed within a temperature range of 10°C to 80°C. The peak accounts for a cooperative transition which is accompanied by a considerable endothermic heat effect. Fractions tp-65 and tp-25 undergo the transition as seen from the corresponding peaks in Figure 6.7. The ts-25 fraction does not precipitate and no transition has been observed for this copolymer. As for ts-65 fraction, its solutions are transparent during heating while a heating peak indicates that ts-65 fraction undergoes coil-to-globule transition without the precipitation of globules.

Random copolymers of N-vinylcaprolactam and N-vinylimidazole synthesized in ethanol were successfully tested as surface nanoreactors in hydrolysis of an ester substrate (see Section 6.3) and therefore it was challenging to test the ts-65 fraction (synthesized in an aqueous medium) in similar experiments. Indeed, the ts-65 fraction can be rather efficient as it is expected to form nearly monomolecular globules capable of adsorbing surface-active substrates. Moreover, such globules should have very high concentration of

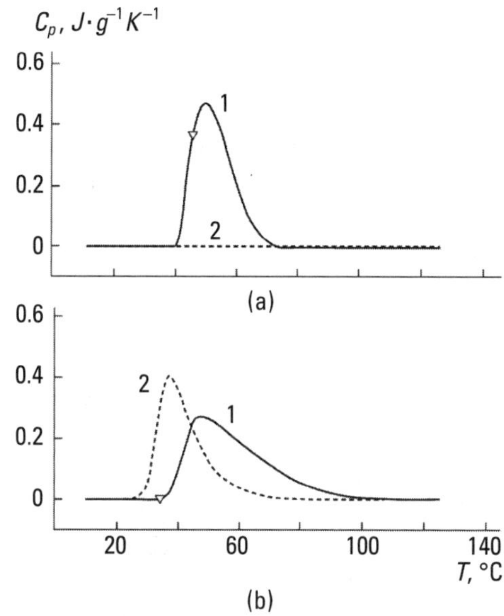

Figure 6.7 Calorimetric properties of thermally precipitating (1) and thermally nonprecipitating (2) fractions obtained at (a) 25°C and (b) 65°C. (*After:* [38].)

N-vinylimidazole groups at the surface (since this is a protein-like copolymer) that is, most hydrophilic groups should be located in the globule shell to prevent aggregation.

It was highly important to elucidate dependence of the catalytic properties on the conformational state which is determined by the temperature of the copolymer solution. Indeed, below 38°C macromolecules of ts-65 fraction exist in a coil conformation. They undergo a transition to a globular protein-like conformation above that temperature. Macromolecules of ts-25 do not change their conformational state upon heating (Figure 6.7). Therefore, to find out whether the globular conformation has an effect on the catalytic properties of protein-like copolymers it was sufficient to compare the catalytic activity of both fractions at lower and higher temperatures.

Ts-65 and ts-25 fractions were prepared by a procedure described in detail previously [37]. Polymerization was carried out in 10% aqueous dimethyl sulfoxide at two temperatures, 25°C (solution polymerization) and 65°C (precipitation polymerization). The overall concentration of the monomers was 0.35 mol/l and the N-vinylcaprolactam to N-vinylimidazole molar ratio was 85:15, these conditions were shown previously [37] to ensure the highest yields of the ts fractions. The ammonium persulfate–N, N, N',

N'-tetramethylethylenediamine redox pair was used as the initiator. The product was purified from monomeric and oligomeric components by dialysis and freeze-dried. Then it was redissolved in water and centrifuged at 65°C and 8,000 rpm for 30 minutes. The supernatant containing the ts fraction was decanted, cooled, and freeze-dried. Some characteristics of these fractions are summarized in the Table 6.3.

Ts-65 and ts-25 fractions were tested in hydrolysis of p-nitrophenyl propionate (Scheme 1) at 25 or 50°C in buffer solutions of N-(2-hydroxyethyl)piperazine-N'-(2-ethanesulfonic acid) (HEPES)–NaOH (0.05 M) at pH 7.3 and 8.2.

Figure 6.8 shows the curves for absorbance variation of p-nitrophenolate-anion at 348 nm during incubation of the solutions (pH 8.2) at 25°C (i.e., below the conformation transition temperature of the ts-65 sample) or at 50°C (*a fortiori* above the conformation transition temperature of this fraction). Decomposition of NPP in the presence of the ts-25 fraction both at 25°C (Figure 6.8, line 2) and at 50°C (Figure 6.8, line 5) was found to proceed at approximately the same rate as the spontaneous hydrolysis. In other words, at the concentration used, the macromolecules of the ts-25 fraction did not show reliably detectable esterolytic activity at 25°C or 50°C. Coil-shaped macromolecules of the ts-65 fraction very slightly catalyzed hydrolysis of NPP at 25°C (like ts-25 sample, Figure 6.8, line 3). However, after the transition of the ts-65-fraction to the protein-like (globular) conformation, the substrate cleavage proceeded approximately 2.3-times faster (Figure 6.8, 6) than the spontaneous hydrolysis or hydrolysis in the presence of ts-25 (50°C, Figure 6.8, lines 4, 5). As the temperature was raised from 25°C to 50°C, the rate of the latter process increased ~5.2-fold (cf. curves 1 and 4, Figure 6.8), while in the presence of ts-65, the reaction rate increased ~12-fold (cf. curves 1 and 6, Figure 6.8). This demonstrates the

Table 6.3
Characteristics of the Polymers Tested in Hydrolysis of p-nitrophenyl Propionate

	M_w	$T_{transition}$	N-vinylimidazole content in % mol, (NMR)
ts-25	120,000	—	38
ts-65	40,000	38	27

Source: [23].

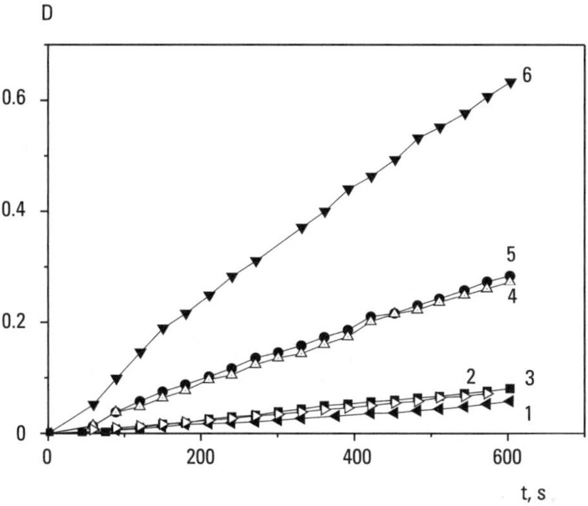

Figure 6.8 Dynamics of liberation of p-nitrophenol in spontaneous hydrolysis of p-nitrophenyl propionate (pH 8.2) at 25 (1) and 50°C (4) and in the hydrolysis in the presence of the copolymer fractions (0.5 μmol/mL relative to the content of N-alkylimidazole groups): ts-25 at 25 (2), and 50°C (5), and ts-65 at 25 (3), and 50°C (6). (*After:* [23].)

catalytic effect caused by specific folding of the copolymer chains where hydrophilic groups of N- alkylimidazole were accumulated in the surface areas of protein-like polymeric particles capable of adsorbing hydrophobic substrate. Depending on the concentration, those are either separate macromolecules or micelles consisting of several macromolecules [39].

Since spontaneous hydrolysis of NPP is rather notable at 50°C and pH 8.2 (Figure 6.8, line 4), it was difficult to elucidate the effects of low concentrations of polymeric catalysts; hence, we also studied hydrolysis in less-alkaline HEPES/NaOH buffer at pH 7.3. The results are presented in Figure 6.9. In this case, a substantial (4.1-fold) increase in the reaction rate was observed in the presence of the protein-like ts-65 fraction even at 25°C (Figure. 6.9, 2) as compared to hydrolysis by the buffer ions (Figure 6.9, 1), while the nonprotein-like ts-25 fraction did not affect hydrolysis of the substrate, and the reaction rate was identical to that of spontaneous hydrolysis (the curve is not shown in Figure 6.9 to avoid superposition of almost identical data). This result suggests that ordered regions may well exist in the structures of macromolecules of the protein-like ts-65 fraction below the conformational transition temperature. This is consistent with the conformational memory-effect predicted theoretically for this type of

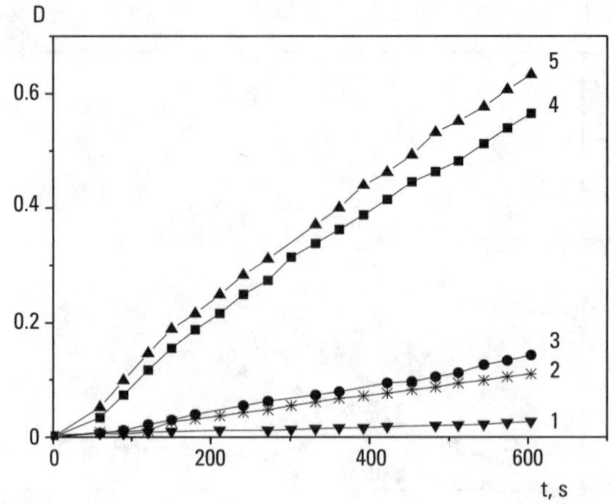

Figure 6.9 Spontaneous hydrolysis of p-nitrophenyl propionate (pH 7.3) at 25 (1) and 50°C (3) and the hydrolysis in the presence of ts-65 fraction (0.5 μmol/mL relative to the content of N-alkylimidazole groups) at (2) 25 and (4) 50°C. Dynamics of the hydrolysis at pH 8.2 and 50°C is also shown for comparison (5). Adapted from [23].

copolymers [29–31, 34]. In addition, the increase of the reaction in the case of ts-65 fraction was markedly greater at pH 7.3 than at pH 8.2. Indeed, spontaneous hydrolysis rate increased 5.4-fold at pH 7.3 (at pH 8.2, 5.2-fold) when the temperature was raised from 25°C to 50°C (cf. curves 1 and 3, Figure 6.9). In the presence of the ts-65 fraction, the reaction rate increased 21.6-fold (curves 1 and 4, Figure 6.9) whereas at pH 8.2, a 12-fold acceleration was observed (see above, the discussion of Figure 6.8). The kinetic curve for hydrolysis by the ts-65 fraction at pH 8.2 (Figure 6.9, 5) is located somewhat higher than that for pH 7.3 (Figure 6.9, 4), which is quite understandable in view of more intensive spontaneous hydrolysis at pH 8.2.

Thus, we have shown that thermoresponsive protein-like copolymers are capable of efficient hydrolysis of ester substrate in globular conformation compared to the copolymers that exist only in unfolded conformation. Such property is most probably connected with the formation of the developed surfaces in which catalytically active imidazole moieties are locally accumulated. Those interfaces are accessible for substrate molecules that can thus concentrate near catalytically active groups.

References

[1] Bronstein L.M., Sidorov S.N., Valetsky P.M. *Russ. Chem. Rev.*, 2004, 73, 501.

[2] Wang G., Kuroda K., Enoki T., Grosberg A., Masamune S., Oya T., Takeoka Y., Tanaka T. *Proc. Natl. Acad. Sci. USA*, 2000, 97, 9861.

[3] Vasilevskaya V.V., Khokhlov A.R. *Macromol. Theory Simul.*, 2002, 11, 623.

[4] Vasilevskaya V.V., Aerov A.A., Khokhlov A.R. *J. Chem. Phys.*, 2004, 120, 9321

[5] Zhao M., Sun L., Crooks R.M. *J. Am. Chem. Soc.* 1998, 120, 4877

[6] Balogh L., Tomalia D.A. *J. Am. Chem. Soc.*, 1998, 120, 7355.

[7] Grohn F., Bauer B.J., Akpalu Y.A., Jackson C.L., Amis E.J. *Macromolecules*, 2000, 33, 6042.

[8] Berezin I.V., Martinek K., Yasimirskii A.K. *Uspekhi khimii* (Russ. Chem. Rev., Russian version), 1973, 10, 1729.

[9] Zakharova L.Ya., Kudryavtseva L.A., Shagidullina R.A. *J. Molec. Liquids*, 2001, 91, 193.

[10] Zakharova L.Ya., Kudryavtseva L.A., Shagidullina R.A., Valeeva F.G. *J. Molec. Liquids*, 2001 94, 79.

[11] Candan F., Pabon M., Anquetil J-Y. *Coll. Surf. A*, 1999, 153, 47.

[12] De la Vega R., Perez-Tajeda P., Lopez-Cornejo P., Sanchez F. *Langmuir*, 2004, 20, 1558.

[13] Liu C.Y., Hu C.C., Hung W.H. J. *Mol. Catal.*, 1996, 106, 67.

[14] Ford W.T. *React. Funct. Polym.*, 1997, 48, 3.

[15] Ford W.T. *React. Funct. Polym.*, 2001, 33, 147.

[16] Vasilevskaya V.V., Aerov A.A., Khokhlov A.R. *Doklady Physical Chemistry*, 2004, 398, 258.

[17] Vasilevskaya V.V., Aerov A.A., Khokhlov A.R. *Colloid Polym. Sci.*, 2006, 284, 459.

[18] Petrovskaya, E. V., Vasilevskaya V. V., Khokhlov, A. R., *Polymer Sci.*, A 2007, 49, 729.

[19] Okhapkin I.M., Bronstein L.M., Makhaeva E.E., Matveeva V.G., Sulman E.M., Sulman M.G., Khokhlov A.R. *Macromolecules*, 2004, 37, 7879.

[20] Okhapkin I.M., Makhaeva E.E., Khokhlov A.R. *Adv. Polym. Sci.*, 2006, 195, 177.

[21] Vasilevskaya V.V., Ablyazov P.N., Khokhlov A.R. *Colloid Journal*, 2007, 69, 265.

[22] Okhapkin I.M., Makhaeva E.E., Khokhlov A.R. *Russ. Chem. Bull.*, 2006.

[23] Lozinskii V.I., Simenel I.A., Khokhlov A.R. *Doklady Chemistry*, 2006, 410, 170.

[24] Kirsh Y.E., Yanul N.A., Kalninsh K.K. *Eur. Polym. J.*, 1999, 35, 305.

[25] Meeussen, F.; Nies, E.; Berghmans, H.; Verbrugghe, S.; Goethals, E.; Du Prez F. *Polymer* 2000, 41, 8597.

[26] Lozinsky V.I., Simenel I.A., Kurskaya E.A., Kulakova V.K., Galaev I.Y., Mattiasson B., Grinberg V.Y., Grinberg N.V., Khokhlov A.R. *Polymer,* 2000, 41, 6507.

[27] Okhapkin I.M., Nasimova I.R., Makhaeva E.E., Khokhlov A.R. *Macromolecules,* 2003, 36, 8130.

[28] Okhapkin I.M., Makhaeva E.E., Khokhlov A.R. *Colloid Polym. Sci.* 2005, 284, 117.

[29] Khokhlov A.R., Khalatur P.G. *Physica A,* 1998, 249, 253.

[30] Khokhlov A.R., Khalatur P.G. *Phys. Rev. Lett.*, 1999, 82, 3456.

[31] Khalatur P.G., Khokhlov A.R. *Adv. Polym. Sci.*, 2006, 195, 1.

[32] Lozinsky V.I. *Adv. Polym. Sci.*, 2006, 196, 87.

[33] Virtanen J., Baron C., Tenhu H. *Macromolecules,* 2000, 33, 336.

[34] Lozinskii V.I., Simenel I.A., Kurskaya E.A., Kulakova V.K., Grinberg V.Ya., Dubovik A.S., Galaev I.Yu., Mattiasson B., Khokhlov A.R. Doklady *Chemistry,* 2000, 375, 273.

[35] Wahlund P.-O., Galaev I.Yu., Kazakov S.A., Lozinsky V.I., Mattiasson B. *Macromol. Biosci.*, 2002, 2, 33.

[36] Siu M.H., Liu H.Y., Zhu X.X., Wu C. *Macromolecules,* 2003, 36, 2107.

[37] Lozinsky V.I., Simenel I.A., Kulakova V.K., Kurskaya E.A., Babushkina T.A., Klimova T.P., Burova T.V., Dubovik A.S., Grinberg V.Ya., Galaev I.Yu., Mattiasson B., Khokhlov A.R. *Macromolecules,* 2003, 36, 7308.

[38] Markova G.D., Vasnev V.A., Keshtov M.L., Vinogradova S.A., Garkusha O.G. *Polym. Sci. A,* 2004, 46, 361.

[39] Lozinsky V.I., Simenel I.A., Semenova M.G., Belyakova L.E., Il'in M.M., Grinberg V.Ya., Dubovik A.S., Khokhlov A.R. *Polym. Sci. A,* 2006, 48, 435.

[40] Perutz M. *Proc. R. Soc. London B,* 1967, 167, 349.

7

Nanoreactors for Enzyme Therapy

Chang-Won Lee and Agnes E. Ostafin

7.1 Enzymes and Disease

In the human body, enzymes produce enantiomerically pure biochemical compounds at ambient temperatures and pressures, often in a single step, something that is very hard to accomplish by conventional organic chemistry. For example, ω-transaminase performs a reversible reaction to transfer amine groups between amino and keto acids (Figure 7.1). The chiral amine products produced by this transformation are biologically active and commercially used in the production of cardiovascular and antihypertensive drugs [1].

Enzymes are indispensable for metabolism, signal transduction and cell regulation, digestion, muscle movement, and intracellular transport. The parallel action of many enzymes leads to the creation of self-regulated, multistep metabolic pathways, in which the product of one enzymatic transformation acts as a substrate for another. Upsetting the balance of this system can lead to a cascade effect in which an entire set of enzymatic reactions are shifted to a new equilibrium. Occasionally, even a slight upregulation of one step can close a metabolic cycle altogether. It is now well-known that

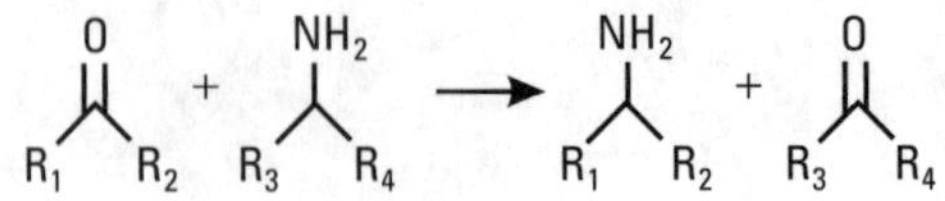

Figure 7.1 ω-transaminase reaction scheme.

enzymatic pathways are hijacked in the pathological mechanisms of virus infections, and serious and even fatal diseases can occur if a critical enzyme is disabled, or the control mechanism of a metabolic pathway is altered. Listed in Table 7.1 are several diseases which arise as a result of a defect in one or more critical enzymes in the human body.

7.2 Enzyme Therapy

Enzyme therapy involves the replacement of a defective or missing enzyme, as well as the use of inhibitors or supplemental substrates to control misregulated enzymes. Because of their unique specificity, enzymes administered in place of synthetic drugs, seem to avoid serious side effects. For example, proteolytic enzymes are widely used as antiinflammatory agents in place of antiinflammatory drugs such as aspirin and Ibuprofen, because they treat the unpleasant inflammatory symptoms without interfering with the production and performance of circulating immune complexes [24]. Roughly 100,000 Americans each year end up hospitalized as a result of immune suppression linked to the use of antiinflammatory drugs [22, 23]. Other examples of effective enzyme therapies include the use of collagenase to help heal dermal ulcers or burns [25], and lysozyme in cases of systemic infection, which along with antibiotics degrade bacterial cell walls [26]. Apart from minor gastrointestinal disturbances, few serious side effects of enzyme therapy have been documented.

Typical sources for highly purified, well-characterized enzymes useful in the treatment of diseases such as those listed in Table 7.2 are the human placenta, spleen, and urine. Genetic engineering methods now permit selective, high-yield production of specific desirable enzymes in controlled cell or tissue culture environments. This reduces the danger of unexpected contamination by viruses, and other bacteria, simplifies the purification process, increases yields, and allows fine tuning of the enzyme performance by substituting, adding, or deleting specific amino acids [27].

Table 7.1
List of Diseases Associated with Enzyme Malfunctions

Name of Disease	Examples of Enzyme Affected	Symptoms	Reference(s)
Diabetes, type 1	glucokinase	Failure to produce insulin, the hormone that "unlocks" or insulin resistance, elevated blood glucose levels, diabetic retinopathy and possible blindness, diabetic neuropathy, heart disease, stroke.	[2, 3]
Fabry disease	α-galactosidase-A	Accumulation of globotriaosylceramidesevere neuropathic, reddish-purple blemishes on their skin, impaired arterial circulation, heart attacks, strokes, kidney failure.	[4]
Gaucher disease	glucocerebrosidase	Accumulation of globotriaosylceramide, pain, fatigue, jaundice, bone damage, anemia, and even death.	[5]
Glycogen storage disease type II (GSD II)	acid α-glucosidase (GAA)	Hypotonia, feeding problems, hepatosplenomegaly, and cardiomyopathy, death.	[6]
Hunter syndrome	iduronate-2-sulfatase	Stature, joint stiffness, coarse facial features, hepato splenomegaly, and progressive mental retardation.	[7]
lymphoyctic leukemias	asparaginase	Accumulation of asparagine.	[8]
Lesch-Nyhan syndrome	hypoxanthine-guanine phosphoribosyl-transferase (HPRT1)	Excess accumulation of uric acid, gout, urate stones, mental problems, muscle weakness.	[9]
Maple syrup urine disease	alpha-ketoacid dehydrogenase (part of BCKD complex)	Accumulation of leucine, isoleucine and valine alpha-ketoacid accumulation in urine, neurodegeneration, death.	[10]
Mucopoly saccharidosis type I (MPS I)	α-L-iduronidase	Hepatosplenomegaly, upper airway obstruction, heart disease, joint stiffness, skeletal dysplasia, corneal clouding, and neurological degeneration, death.	[11], [12]
Niemann–Pick disease	sphingomyelinase	Accumulation of sphingomyelin.	[13–15]

Table 7.1 (continued)

Name of Disease	Examples of Enzyme Affected	Symptoms	Reference(s)
Obesity	lipoprotein lipase stearoyl-CoA desaturase-1 (SCD-1) 11beta-hydroxysteroid dehydrogenase type 1 (11beta-HSD1)	Leptin deficiency, increased storage of fat.	[16, 17]
Pheny-lketonuria	phenylalanine hydroxylase	Buildup of phenylalanine.	[18]
Refsum disease	PAHX - phytanoyl coenzyme A dydroxylase	Accumulation of phytol, peripheral neuropathy, ataxia, retinitis pigmentosa, bone and skin changes.	[19]
Tay-Sachs disease	enzyme β-hexosaminidase A	Accumulation of GM2 ganglioside, paralysis, dementia, blindness and early death to a chronic adult form that exhibits neuron dysfunction and psychosis.	[20], [21]

7.2.1 Intravenous Administration and Chemical Modification of Enzymes for Therapeutic Use

The activity of enzymes injected directly into plasma decreases exponentially with time after injection due to denaturation and proteolysis. The rate of denaturation and proteolysis depends on molecular size, charge, surface functional groups, and glycosylation. Strategies for prolonging the effectiveness of enzyme therapy include: intravenous injection, the use of chemically modified enzymes that prolong circulation half-life, antibody, and viral vector targeting to increase retention in specific cells and tissues, and micro- or nanoencapsulation to protect the enzyme until it is needed or reaches a target site.

In addition to structural and chemical changes, intravenously administered enzymes can be blocked by inhibitors present in the blood. These are usually organic or peptide molecules that either, resemble the desired substrate and compete for interaction, or occlude the enzyme's active site. Most healthy organisms manage to control the level of inhibitors by a complex balance of production and elimination reactions. In the case of enzymes externally administered to replace a missing or defective enzyme, the patient may also not have a control system for the regulation of inhibitors for that

Table 7.2
List of Enzyme-Carrying Nanoparticles

Method/ Location	Material	Size (nm)	Enzyme	Reference(s)
Emulsion/ entrapment	poly(lactide-co-glycolide)	196-226	L-asparaginase	[63]
Emulsion/ entrapment	polyacrylamide	31	Horseradish peroxidase	[64]
Emulsion/ surface	Phospholipid shell with polystyrene	300–500	Acetylcholinesterase, choline oxidase, horseradish peroxidase	[65]
Emulsion/ entrapment	gold	50–95	Horseradish peroxidase	[66]
Emulsion/ entrapment	silicate	<100	Horseradish peroxidase	[69]
Emulsion/ entrapment	silica coated sliver	40	Horseradish peroxidase	[70]
Emulsion/ surface	silicate	30–100	Glucose oxidase	[71]
Emulsion/ surface	silica coated Fe_3O_4	9.1	β-lactamase	[72]
Emulsion/ surface	Polystyrene-b-polyacrylate coated Fe_2O_3	50	T4 DNA ligase	[73]
Biosilification/ entrap	R5 protein with silicate	500	Horseradish peroxidase	[74]
Dendrimer/ entrapment	silica coated PAMAM dendrimer	162	Horseradish peroxidase	[75]
Gold reduction/ surface	gold	30	Glucose oxidase	[76]
Gold reduction/ surface	gold	10	Catalase	[78]
Gold reduction/ surface	gold	13	Esterase	[79]
Coating/ entrapment	silicate	4-8	Chymotrypsin, trypsin	[80]

particular enzyme. Since this would lower the therapeutic effectiveness, this must be taken into account when formulation is designed. Similarly, the patient may experience an allergic reaction to the foreign enzyme stimulating accelerated denaturation and removal from the body [28].

Direct chemical modification of enzymes in order to increase residence time in the body, and protect against proteolysis, inhibitors and immune recognition has shown some success. For example, chemical modification of phenylalanine ammonia lyase (PAL) by pegylation produces a protected form of PAL that possesses better specific activity, prolonged half-life, and reduced immunogenicity in vivo. Subcutaneous administration of pegylated PAL to PKU mice had the desired metabolic response (prolonged reduction in blood and brain Phe levels) with greatly attenuated immunogenicity [29]. Catalase, an enzyme which degrades hydrogen peroxide, a major product in the metastasis of cancer cells, is another example of an enzyme therapeutic whose retention time and inhibition of pulmonary metastases is significantly improved by pegylation [30].

7.2.2 Antibody and Viral Vector Targeting of Enzyme Therapies

While pegylation improves the durability of therapeutic enzymes in the body, it has little effect on targeting the enzyme to specific cells or tissues which may be most affected by enzyme malfunction. Enzymes administered directly to the bloodstream may not have access to certain tissues/cells because of their physical properties, or the differential expression of receptors in certain cell types. Such environments include the brain, bones, lymph nodes, nervous system, and other differentiated cells with specific functions. For instance, β-hexosaminidase administered intravenously to a patient with gangliosidosis appeared in the liver, but not in cerebrospinal fluid or biopsied brain tissue [31]. Direct administration of the enzyme into the cerebrospinal fluid also did not improve the accumulation of enzyme in the brain. In contrast, antibody-conjugated [34] β-galactosidase was able to enter the brain in significant amounts [37]. However, enzymes conjugated to antibodies also increase the problem of immunogenicity and require the use of immunosuppressive agents, which have their own dangers. Lipophilic nanoparticles less than 500 Da in size can diffuse across the blood-brain barrier by direct permeation [32], but may be easily removed by efflux transporters [33].

Gene therapy may be another way to deliver enzymes to desired cells or tissues [35, 36]. Instead of delivering enzymes or proteins to the target intravenously or by subcutaneous injection, a gene which encodes the structure of the desired enzyme is delivered to the target cell by a virus where its expression is induced. Viral vectors used in gene therapy are typically adenoviral vectors and are designed to integrate the desired gene sequence into the genome of target cells from where it can be expressed to produce functional enzymes using the cell's normal processes. Gene delivery does little to protect the long-term viability of enzymes product or to reduce its immunogenicity. Its main advantage is that it creates a continuous and permanent source of new enzymes in the targeted tissues. There are many other unsolved problems associated with gene delivery including how to ensure delivery to specific target cells, achieving site-specific integration into the host chromosome, controlling the expression, and controlling the immune response of the patients if they are allergic to the enzyme product [39]. Uncontrolled, these problems can lead to serious side effects, and fatality [38].

7.2.3 Microreactor Immobilization of Enzyme Therapies

A conceptually simpler way to deliver enzyme therapeutics is to immobilize them in or on micro- or nanoparticles. Because these particles retain the chemical activity of the enzyme they are often referred to as micro- or nanoreactors. This strategy seems to increase the stability of easily denatured enzymes, obscures them from immune recognition, and prolongs their circulatory half-life. In the first generation of such products, enzymes were simply attached to surfaces of polymeric or inorganic particles via physisorption, covalent bonding or ionic interactions. Although surface immobilization techniques are relatively easy, enzymes lose activity due to a random orientation of immobilization which blocks or limits the accessibility to enzyme's active site. Depending on the particle's surface chemistry, the flexibility and structure of enzymes can be compromised by physical binding of the part of the enzyme to the particle surface. Since the enzyme remains continually exposed to its surroundings, changes in chemical composition and air drying can still lead to denaturation and loss of activity. Nevertheless, more than 5,000 papers and patents have been published on enzyme immobilization technologies [40] since Nelson and his colleagues first immobilized invertase on $Al(OH)_3$ by simple adsorption [41]. Polymer beads such as Eupergit C and chitosan [42, 43] and γ-PGA-graft-L-PAE nanoparticles [44] have been widely used for the surface immobilization of enzymes.

Entrapment of enzymes inside polymer matrices can be achieved by mixing monomers and enzymes together and then crosslinking the polymer to trap enzymes inside gaps in the network. The ability of this approach to maintain the enzymatic activity was first demonstrated in slab gels of inorganic silica [46]. Other polymer gels used for entrapping enzymes include polyacrylamide [47, 49], pHEMA (polyhydroxyethylmethacrylate) [48], silicate [45], dextran [50], alginate [51], and PU (polyurethane) [52]. Common characteristics of these polymer systems are their hydrophilicity, mild polymerization process conditions such as neutral pH and room temperature, few reaction side products to inhibit or denature enzymes, and enough porosity to allow substrates and products to pass while still keeping enzymes trapped inside.

However, slab gels are not ideal for the delivery of enzyme therapeutics, even in the form of long-term implants, since their macroscopic size creates intrinsic mass-transport bottlenecks and results in an overall slowed enzymatic reaction rate. One of only a few examples using slab gels for enzyme therapy is the entrapment of cytosine deaminase in an epoxy-acrylic resin/urethane prepolymer hydrogel implant for cancer chemotherapy [53]. Another example is millimeter-sized poly-(lactic-co-glycolic acid) slab gel implants used to achieve sustained delivery of inhibitors of aldose reductase in a rat model [54]. A second problem with the slab gel size, is that there is a potential for foreign body and inflammatory responses from the body [55]. From a therapeutic perspective, the location of the implant is an important consideration. In many cases, the targets that need to be treated are spread throughout the body or located in inaccessible spots making it difficult to implant a relatively large-sized material.

An elegant way to protect enzymes for therapeutic function while minimizing transport bottlenecks is to use hollow microparticles, called microreactors [36, 45]. Microreactors are designed specifically to be hosts for chemical reactions, and encapsulated enzymes are held in a drop of fluid separated from their surroundings by a semipermeable layer of a polymer, lipid, or inorganic substance. The coating layer must be semipermeable to allow the substrate access to the enzyme, but still limit the accessibility of other substances. The diffusion of the substrate, and to a greater or lesser extent that of the product, establishes the immobilized enzyme's overall activity. If the porosity of the encapsulating layer is allowed to become too large then the enzyme can leach out.

The advantage of using hollow microreactors can be seen by considering Fick's law (7.1). The velocity of diffusion of an enzyme substrate is:

$$v_d = D_e \cdot \frac{F}{r} \cdot \Delta S \tag{7.1}$$

Where D_e is diffusion constant (m^2/s), F is diffusion surface area (m^2), r is diffusion distance (m) and ΔS is difference between substrate concentration at the beginning (bulk concentration) and the end (concentration at the center of the bead or slab) of the diffusion distance (mM/m^3). For a given solid matrix and enzyme, reducing the diffusion distance, *r* or increasing diffusion surface area *F*, improves substrate flux and lead to more efficient reaction. For a microreactor consisting of a hollow liquid-filled core surrounded by a nanometers-thin coating, *r* is extremely small. Only nonideal effects inside the microreactors, such as increased viscosities and electrostatic interactions near the interfaces can reduce the velocity of substrate uptake significantly.

For a solid microparticle enzymatic reactions can appear to follow a range of kinetics ranging from Michaelis-Menten at the surface to nearly first-order kinetics on the interior. The rate is limited by the concentration of substrate at various depths within the particle. This can be expressed via a dimensionless quantity called the Theile modulus **Φ** [56],

$$\Phi = \frac{r}{3} \cdot \sqrt{\frac{V_{\max}}{K_m \cdot D_e}} \tag{7.2}$$

where *r* is the diffusion distance (*m*), V_{max} and K_m are the Michaelis-Menten kinetics reaction constants and D_e is diffusion constant (m^2/s). Only when the Thiele modulus is close to zero is the substrate/product distribution uniform throughout the particle. In the case of hollow microreactors, once steady state has been achieved, there should be no significant gradient in the substrate or product concentration inside the liquid core of this type of microparticle. The diffusion coefficient of the substrate inside the nanoreactor should be close to that in the solution yielding maximum reaction efficiency. The thickness of the microreactor coating is the only factor which controls the reaction velocity (permeability). A lowered permeability, and a fast enzymatic reaction results in a low-average concentration of substrate in the microreactor and a lower turnover rate. In contrast, an improved permeability leads to a higher average concentration of substrate and faster turnover. In such a case the effectiveness factor, η, which is the ratio of the apparent reaction rate to that evaluated in dilute solution should be close to one. The relationship between the Theile modulus and effectiveness factor is typically illustrated by graphs such as that shown in Figure 7.2.

7.2.4 Nanoreactor Immobilization of Enzyme Therapies

Microreactors are relatively large structures similar in size to pathogenic cells and so may become trapped in tissue capillaries or recognized by the immune

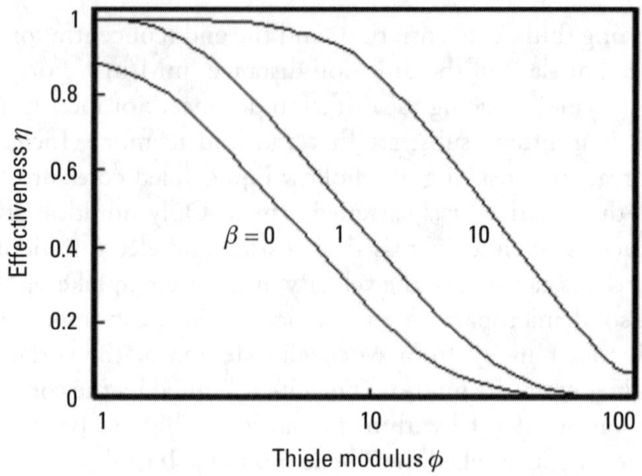

Figure 7.2 Typical relationship of effectiveness factor versus Thiele modulus for different concentration, β.

system as a foreign cell. As a result, nanoreactor immobilization of therapeutic enzymes is of increasing interest. Nanoparticles and nanoreactors can be easily transported into cells, and even the nucleus of cells, with either passive diffusion [57] or active transport using signaling molecules [58]. Unlike much larger microparticles, the surface charge of nanoparticles assists in shielding van der Waals forces, and minimizing surface reactions that would otherwise neutralize the particles and cause them to aggregate. The smaller the size, the more likely they will behave as molecules whose Brownian motion is energetic enough to overcome gravity so that the suspension remains well-dispersed [59]. Particle settling velocity, v, can be expressed by Stoke's law:

$$v = \frac{d^2 g(\rho_s - \rho_l)}{18\mu_l} \tag{7.3}$$

where g is gravitation acceleration (9.8 m/s), ρ_l is liquid density (997 kg/m^3 for water at 25°C), and μ_l is the viscosity of liquid water (0.00089 Pa/s). If we assume the particle's density, ñ$_s$, to be 1,200 kg/m^3 (a typical polymer density), we can calculate the settling velocity, v, of the nanoparticles in water. For a 1-nm diameter particle, the settling velocity as 0.00124 (nm/s), while for 10-nm particles, it is 0.124 (nm/s) and for 1-μm particles, it is

1,240 (nm/s). By Einstein's fluctuation-dissipation theory, we can express the average Brownian displacement x in time t as:

$$x = \sqrt{\frac{2k_B Tt}{\pi \mu d}} \tag{7.4}$$

where k_B is the Boltzman constant (1.38×10^{-23} J/K), and T is temperature in Kelvin. If the water temperature is 25°C, the Brownian displacement of 1-nm diameter nanoparticles is 54,250 nm/s, for 10 nm, it is 17,155 nm/s and for 1 μm, it is 1,716 nm/s. These values are well above those for the gravitational settling velocity, so the nanoparticles can stay uniformly dispersed in liquids.

Enzymatic nanoreactors, both solid and hollow, have been produced using a wide range of materials and synthesis procedures, the more notable of which are listed in Table 7.2.

7.2.4.1 Emulsions

Emulsion methods are one of the most widely used methods for preparing nanoreactors. If enough energy is provided to a mixture of two immiscible solvents, the solvents will disperse as small droplets inside the other. Hardening or stabilization of the emulsion using emulsifiers is required to make a nanoparticle. The scheme for this process is illustrated in Figure 7.3.

Quintanar-Guerrero et al. showed that larger amounts of emulsifiers can reduce the size of nanoparticles well into 100-nm in diameter range [60]. The stirring rate [61] and pressure [62] are also important parameters.

Enzymes immobilized in emulsion-prepared nanoreactors appear to retain much of their native activity. In 1998, Gasper et al. entrapped L-asparaginase into poly(lactide-co-glycolide) (PLG) nanoreactors prepared by water/oil/water emulsion methods [63]. They showed that higher molecular weight PLG had a higher loading and slower release of protein from nanoparticles than did lower molecular weight PLG nanoreactors. Later, Poulsen et al. showed that horseradish peroxidase (HRP) embedded in polyacrylamide nanoparticles prepared by the emulsion method retained 75% of its activity, as measured by the response of coentrapped fluorescent dyes to the reactive oxygen species generated by this enzyme [64].

Multistep enzymatic reactions are more efficient using nanoreactors. A series of enzymes were immobilized onto 360-nm diameter phospholipid-coated polystyrene nanoparticles prepared by emulsion methods. Acetylcholinesterase, choline oxidase, and horseradish peroxidase were

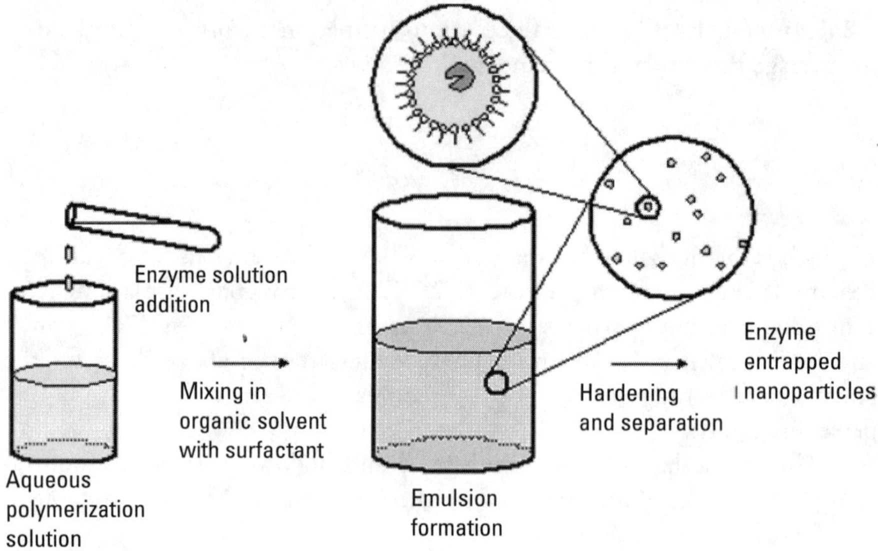

Figure 7.3 Typical scheme for the production of enzyme-entrapped nanoparticles by an emulsion method.

coimmobilized on the nanoparticle and series of enzyme reactions carried out to convert the primary substrate, acetylcholine chloride to hydrogen peroxide which was detected by the oxidization of tetramethylbenzidine [65]. The sequential reactions on the nanoparticles were found to be significantly faster than that of the enzyme mixture in solution, and attributed to the close localization of key enzymes for the reactions.

Even though emulsion methods are one of the most widely used methods to prepare enzyme nanoreactors, the method has some basic drawbacks. Exposure of enzymes to organic solvents or emulsifier during the formation process may lead to denaturation of the weaker enzymes and aggregation between the nanoparticles during the polymerization can reduce the surface accessibility.

7.2.4.2 Gold

Gold nanoparticles [77] have also been used to construct nanoreactors for enzymatic immobilization. Gold is generally thought to be safe to use in living organisms, since toxicity is related with reactive and soluble Au^{+} and Au^{3+} species [82]. The prevalence of allergic reactions triggered by gold affect up to 13% of population.

Enzymes are attached to the gold nanoparticles surface by covalent surface immobilization, or adsorption. Li and coworkers [76] reported that the immobilized glucose oxidase showed improved thermal stability compared to free enzyme. Other examples include catalase [78] and hexa-arginine tagged esterase [79] which retained about 60% of their activity.

Kumar et al. prepared hollow gold nanoreactors with a mean diameter around 50 to 95 nm using gold-coated silver nanoparticles prepared using a reverse micelle system, by leaching out a core silver chloride (AgCl) and leaving an outer shell of gold [66]. They showed that they can effectively encapsulate HRP without compromise in activity in these particles. Small substrates which were able to pass through the pores of the particle were used to measure the entrapped enzyme's Michaelis-Menten kinetics. As expected based on the sieving effect of the gold lattice, large substrates did not induce activity. However, the enzyme kinetics study using smaller substrates showed that the entrapped enzymes were still active, although they seemed to have less affinity for the substrate compared to the free enzymes. This was attributed to the constrained diffusion of substrates inside the nanoreactors, unlike the case of hollow microreactors where the liquid volume is large enough and local concentration of enzyme low enough that steric problems are less important and normal kinetic prevail.

7.2.4.3 Silica

Silica nanoreactors are prepared using sol-gel methods [67–69], using mildly acidic or basic conditions and silica precursors such as tetramethoxysilane (TMOS) or tetraethoxysilane (TEOS). Methanol or ethanol, respectively, is a side product of the formation of a Si-O-Si backbone. Silicate is generally considered a nontoxic compound for human health and no adverse effects are observed at exposure levels of 50,000 ppm [82]. Typically, enzymes are mixed directly with the aqueous silica polymerization solution prior to its polymerization and become entrapped inside the network during gel formation. By performing the hardening reaction in aqueous phase at neutral pH and minimizing the use of denaturing solvents like n-hexane, silica encapsulated enzymes such as HRP are able to maintain much of their native activity. Glucose oxidase immobilized on the surface of silicate nanoreactors with varying nanosized diameters between 30 to 100 nm had better residual activity on smaller nanoparticles suggesting that the surfaces of smaller nanoparticles had less of a denaturing effect on the enzymes [71].

Tsang et al. prepared silica-coated-magnetic nanoreactors in a water-in-oil emulsion [72]. Tetraethoxysilane (TEOS) was used to form porous silica shells over superparamagnetic Fe_3O_4 nanoparticles 9.10 ± 0.39 nm in diameter, that was further functionalized with aminopropyltriethoxysilane (APS) to provide amine groups on the surface for enzyme crosslinking. β-lactamase covalently attached by glutaraldehyde exhibited Michaelis-Menten kinetics parameters as good as those of the free enzyme. Similarly, Herdt et al. reported magnetic Fe_2O_3 nanoparticles encapsulated within crosslinked polystyrene-*block*-polyacrylate copolymer micelle shells [84]. The 50-nm particles surface were functionalized with Cu^{2+}-iminodiacetic acid and His-tagged T4 DNA ligase successfully attached.

Another interesting approach to prepare silica nanoreactors is to utilize templating compounds such as the silaffin protein R5 (H_2N-SSKKSGSYSGSKGSKRRIL-COOH) from *Cylindrotheca fusiformis* to catalyze the formation of silica. Naik and colleagues used this natural product to produce enzymatic nanoreactors [74]. To gain more control over nanoreactor size and homogeneity, dendrimers, specialized self-assembling-branched polymers can be utilized as a template for silification. Unlike unbranched symmetric polymers which aggregate randomly by van der Waals and sometime electrostatic interactions, or the R5 peptide, the branched polymers used to make dendrimers are asymmetric and their assembly is directional from the inside out. Miller et al. described a method to prepare HRP-filled silica nanoparticles by using amine-terminated dendrimers (Polyamidoamine, PAMAM) as a template [76]. The amine-terminated dendrimer acted as a catalyst for the condensation reaction of $Si(OH)_4$ to create silica coating which also trapped the enzymes inside the network. On average several dendrimers and enzymes were encapsulated in one 162 ± 92-nm diameter particle. Although this approach is similar to the silica sol-gel methods mentioned above except for the dendrimer, the enzyme's specific activity decreased about 50% during the process.

Silica sol-gel chemistry has also been used to form hollow silica nanoreactors encapsulating HRP [70] and alcohol dehydrogenase [82]. In the case of alcohol dehydrogenase gold nanoparticles were used as a platform to first bind the enzyme, which was then overcoated with a layer of silicate. To enhance the survival of the enzyme during this process the pH of the reaction was shifted from that of the optimum pH for enzyme activity.

The smallest possible enzyme nanoreactor reported to date would be a single enzyme nanoparticle (SEN) created by Kim et al. from Pacific Northwest National Laboratory [80]. Proteases (α-chymotrypsin and trypsin) were coated with a barrier layer of silicate applied in two different polymerization steps. The enzyme was first modified with acryloyl chloride to introduce polymerizable vinyl units on the surface of the enzyme and then polymerized with monomers containing both vinyl and trimethoxysilane groups. Once the radical polymerization by vinyl groups was complete, the second condensation reaction by trimethoxysilane groups was used to cover the enzyme surface

SENs made of chymotrypsin and trypsin retained 38% to 73% of their original activity comparing to the free enzymes, along with much improved thermal and proteolytic stability. For example, after incubation in 30°C buffer solution for two days, the activity of free chymotrypsin was less than 10% while SEN-chymotrypsin did not show any decrease of activity.

7.3 Summary

Many strategies have been developed to overcome the intrinsic problems associated with therapeutically administered enzymes. The latest include the use of micro- and nanoreactors as carrier vessels for enzymes. By choosing a biocompatible, chemically inert substance as the matrix, such as biodegradable polymers, gold or silicate, a versatile therapeutic carrier can be designed that retains much of the activity of the native enzyme. The matrix can also be used as a platform for the attachment of secondary compounds, such as inhibitors or regulators of enzymatic pathways, and site-directing units such as antibodies to direct their carriers to affected tissues, or prolong their circulatory half-life. The small size of nanoreactors minimizes transport bottlenecks, leads to more stable suspensions, and allows uptake by cells via endocytosis. While there are currently no clinically used nanoreactor formulations for enzymatic therapy, this technology is definitely on the horizon.

References

[1] Shin, J. S.; Kim, B. G. *Biotechnol. Bioeng.* 1998, *60*, 534–540.

[2] Matschinsky, F. M.; Magnuson, M. A.; Zelent, D.; Jetton, T. L.; Doliba, N.; Han, Y.; Taub, R.; Grimsby, J. *Diabetes* 2006, *55*, 1–12.

[3] Montane, J.; Mas, A.; Chillon, M.; Anguela, X. M.; Munoz, S.; Riu, E.; Otaegui, P. J.; Bosch, F. *Mol. Ther. 2004, 9*, S59–S59.

[4] Schiffmann, R.; Kopp, J. B.; Austin, H. A., 3rd; Sabnis, S.; Moore, D. F.; Weibel, T.; Balow, J. E.; Brady, R. O. *JAMA* 2001, *285*, 2743–2749.

[5] Neufeld, E.; Maunzer, J. *The Mucopolysaccharidoses*; McGraw-Hill, 2001.

[6] Hirschhorn, R.; Reuser, A. *Glycogen Storage Disease Type II*; McGraw-Hill, 2001.

[7] McBride, K. L. *Expert Review of Endocrinology and Metabolism* 2007, *2*, 19–26.

[8] Kantarjian, H. M. *Am. J. Med.* 1994, *97*, 176–184.

[9] Nyhan, W. L. *J Hist. Neurosci.* 2005, *14*, 1–10.

[10] Mitsubuchi, H.; Owada, M.; Endo, F. *J. Nutr.* 2005, *135*, 1565S–1570S.

[11] Ross, C. J.; Bastedo, L.; Maier, S. A.; Sands, M. S.; Chang, P. L. *Hum. Gene Ther.* 2000, *11*, 2117–2127.

[12] Miebach, E. *Acta Paediatr. Suppl.* 2005, *94*, 58–60; discussion 57.

[13] Daloze, P.; Delvin, E. E.; Glorieux, F. H.; Corman, J. L.; Bettez, P.; Toussi, T. *Am. J. Med. Genet.* 1977, *1*, 229–239.

[14] Miranda, S. R.; He, X.; Simonaro, C. M.; Gatt, S.; Dagan, A.; Desnick, R. J.; Schuchman, E. H. *Faseb J* 2000, *14*, 1988–1995.

[15] Patterson, M. C. *Neurologist* 2003, *9*, 301–310.

[16] Tomlinson, J. W.; Stewart, P. M. *Nat. Clin. Pract. Endocrinol. Metab.* 2005, *1*, 92–99.

[17] Muoio, D. M.; Newgard, C. B. *Annu. Rev. Biochem.* 2006, *75*, 367–401.

[18] MacDonald, M. J.; D'Cunha, G. B. *Biochem. Cell Biol.* 2007, *85*, 273–282.

[19] Jansen, G. A.; Waterham, H. R.; Wanders, R. J. *Hum. Mutat.* 2004, *23*, 209–218.

[20] Ross, C. J.; Ralph, M.; Chang, P. L. *Exp. Neurol.* 2000, *166*, 276–286.

[21] Lemieux, M. J.; Mark, B. L.; Cherney, M. M.; Withers, S. G.; Mahuran, D. J.; James, M. N. *J Mol Biol* 2006, *359*, 913–929.

[22] *Regular Use of Pain Relievers Can Have Dangerous Results*, American Medical Association 1997.

[23] *Wall Street Journal* 1999; Vol. 19 April.

[24] Wrba, H.; Pecher, O. *Enzymes: A Drug of the Future*; Eco Med, 1993.

[25] Li, Y.; Li, G. Q.; Lin, C. M.; Cai, X. N. *Journal of Dermatological Science* 2005, *37*, 58–60.

[26] Masschalck, B.; Deckers, D.; Michiels, C. W. *J. Food Prot.* 2002, *65*, 1916–1923.

[27] Lamla, T.; Mammeri, K.; Erdmann, V. A. *Acta Biochim. Pol.* 2001, *48*, 453–465.

[28] Linthorst, G. E.; Hollak, C. E. M.; Donker-Koopman, W. E.; Strijland, A.; Aerts, J. M. F. G. *Kidney Int.* 2004, *66*, 1589–1595.

[29] Gamez, A.; Sarkissian, C. N.; Wang, L.; Kim, W.; Straub, M.; Patch, M. G.; Chen, L.; Striepeke, S.; Fitzpatrick, P.; Lemontt, J. F.; O'Neill, C.; Scriver, C. R.; Stevens, R. C. *Mol. Ther.* 2005, *11*, 986–989.

[30] Nishikawa, M.; Hyoudou, K.; Kobayashi, Y.; Umeyama, Y.; Takakura, Y.; Hashida, M. *J. Control Release* 2005, *109*, 101–107.

[31] Itakura, T.; Kuroki, A.; Ishibashi, Y.; Tsuji, D.; Kawashita, E.; Higashine, Y.; Sakuraba, H.; Yamanaka, S.; Itoh, K. *Biol. Pharm. Bull.* 2006, *29*, 1564–1569.

[32] Pardridge, W. M. *J. Neurochem* 1998, *70*, 1781–1792.

[33] Tsuji, A.; Tamai, I. I. *Adv. Drug Deliv. Rev.* 1999, *36*, 277–290.

[34] Shuvaev, V. V.; Christofidou-Solomidou, M.; Scherpereel, A.; Simone, E.; Arguiri, E.; Tliba, S.; Pick, J.; Kennel, S.; Albelda, S. M.; Muzykantov, V. R. *J. Control Release* 2007, *118*, 235–244.

[35] Campos, S. K.; Barry, M. A. *Curr. Gene Ther.* 2007, *7*, 189–204.

[36] Borm, P. J.; Schins, R. P.; Albrecht, C. *Int. J. Cancer* 2004, *110*, 3–14.

[37] Zhang, Y.; Pardridge, W. M. *J. Pharmacol. Exp. Ther.* 2005, *313*, 1075–1081.

[38] Lehrman, S. *Nature* 1999, *401*, 517–518.

[39] Verma, I. M.; Somia, N. *Nature* 1997, *389*, 239–242.

[40] Cao, L. *Carrier-Bound Immobilized Enzymes*, Wiley-VCH, 2005.

[41] Nelson, J. M.; Griffin, E. G. *J. Am. Chem. Soc.* 1916, *38*, 1109–1115.

[42] Katchalski-Katzir, E.; Kraemer, D. M. *Journal of Molecular Catalysis B: Enzymatic* 2000, *10*, 157–176.

[43] Yi, S.-S.; Lee, C.-W.; Kim, J.; Kyung, D.; Kim, B.-G.; Lee, Y.-S. *Process Biochemistry* 2007, *42*, 895–898.

[44] Akagi, T.; Kaneko, T.; Kida, T.; Akashi, M. *J. Biomater. Sci. Polym. Ed.* 2006, *17*, 875–892.

[45] Lee, C.-W.; Yi, S.-S.; Kim, J.; Lee, Y.-S.; Kim, B. G. *Biotechnol. Bioprocess. Eng.* 2006, *11*, 277–281.

[46] Dickey, F. H. *J. Phys. Chem.* 1955, *59*, 695–707.

[47] Bernfeld, P.; Wan, J. *Science* 1963, *142*, 678–679.

[48] Kok, F. N.; Bozoglu, F.; Hasirci, V. *J. Biomater. Sci. Polym. Ed.* 2001, *12*, 1161–1176.

[49] Tumturk, H.; Karaca, N.; Demirel, G.; Sahin, F. *Int. J. Biol. Macromol.* 2007, *40*, 281–285.

[50] Franssen, O.; Stenekes, R. J.; Hennink, W. E. *J. Control Release* 1999, *59*, 219–228.

[51] Betigeri, S. S.; Neau, S. H. *Biomaterials* 2002, *23*, 3627–3636.

[52] Hu, Z. C.; Korus, R. A.; Stormo, K. E. *Appl. Microbiol. Biotechnol.* 1993, *39*, 289–295.

[53] Katsuragi, T.; Sakai, T.; Tonomura, K. *Appl. Biochem. Biotechnol.* 1987, *16*, 61–69.

[54] Aukunuru, J. V.; Sunkara, G.; Ayalasomayajula, S. P.; DeRuiter, J.; Clark, R. C.; Kompella, U. B. *Pharm. Res.* 2002, *19*, 278–285.

[55] Schlosser, M.; Wilhelm, L.; Urban, G.; Ziegler, B.; Ziegler, M.; Zippel, R. *J. Biomed. Mater. Res.* 2002, *61*, 450–457.

[56] Hartmeier, W. *Immobilized Biocatalysts*; Springer-Verlag, 1988.

[57] Panyam, J.; Lof, J.; O'Leary, E.; Labhasetwar, V. *J Drug Target* 2002, *10*, 515–523.

[58] Moghimi, S. M.; Hunter, A. C. *Crit Rev Ther Drug Carrier Syst* 2001, *18*, 527–550.

[59] Gupta, R.; Kompella, U. B. *Nanoparticle Technology for Drug Delivery*; Taylor & Francis, 2006; Vol. 159.

[60] Quintanar-Guerrero, D.; Fessi, H.; Allemann, E.; Doelker, E. *International Journal of Pharmaceutics* 1996, *143*, 133–141.

[61] Cegnar, M.; Kos, J.; Kristl, J. *Eur. J. Pharm. Sci.* 2004, *22*, 357–364.

[62] Yoncheva, K.; Vandervoort, J.; Ludwig, A. *J. Microencapsul.* 2003, *20*, 449–458.

[63] Gasper, M. M.; Blanco, D.; Cruz, M. E.; Alonso, M. J. *J. Control Release* 1998, *52*, 53–62.

[64] Poulsen, A. K.; Scharff-Poulsen, A. M.; Olsen, L. F. *Anal. Biochem.* 2007, *366*, 29–36.

[65] Watanabe, J.; Ishihara, K. *Biomacromolecules* 2006, *7*, 171–175.

[66] Kumar, R.; Maitra, A. N.; Patanjali, P. K.; Sharma, P. *Biomaterials* 2005, *26*, 6743–6753.

[67] Gill, I.; Ballesteros, A. *Trends Biotechnol.* 2000, *18*, 469–479.

[68] Gill, I.; Ballesteros, A. *Trends Biotechnol.* 2000, *18*, 282–296.

[69] Cellesi, F.; Tirelli, N. *Colloids and Surfaces A: Physicochemical and Engineering Aspects* 2006, *288*, 52–61.

[70] Sharma, R. K.; Das, S.; Maitra, A. *J. Colloid Interface Sci.* 2005, *284*, 358–361.

[71] Sun, Y.; Yan, F.; Yang, W.; Zhao, S.; Yang, W.; Sun, C. *Anal. Bioanal. Chem.* 2007, *387*, 1565–1572.

[72] Tsang, S. C.; Yu, C. H.; Gao, X.; Tam, K. *J. Phys. Chem. B* 2006, *110*, 16914–16922.

[73] Herdt, A. R.; Kim, B. S.; Taton, T. A. *Bioconjug. Chem.* 2007, *18*, 183–189.

[74] Naik, R. R.; Whitlock, P. W.; Rodriguez, F.; Brott, L. L.; Glawe, D. D.; Clarson, S. J.; Stone, M. O. *Chemical Communications* 2003, 238–239.

[75] Miller, S. A.; Hong, E. D.; Wright, D. *Macromol. Biosci.* 2006, *6*, 839–845.

[76] Li, D.; He, Q.; Cui, Y.; Duan, L.; Li, J. *Biochem. Biophys. Res. Commun.* 2007, *355*, 488–493.

[77] Link, S.; El-Sayed, M. A. *J. Phys. Chem. B* 1999, *103*, 8410–8426.

[78] Patel, N.; Davies, M. C.; Hartshorne, M.; Heaton, R. J.; Roberts, C. J.; Tendler, S. J. B.; Williams, P. M. *Langmuir* 1997, *13*, 6485–6490.

[79] Ha, T. H.; Jeong, J. Y.; Chung, B. H. *Chemical Communications* 2005, 3959–3961.

[80] Kim, J.; Grate, J. W. *Nano Lett.* 2003, *3*, 1219–1222.

[81] Martin, K. R. *J. Nutr. Health Aging* 2007, *11*, 94–97.

[82] Wang, Q.; Chen, X.; Mizukami, H.; Miesel, D.; Ostafin A. E. SPIE Proceedings, San Jose, CA, January 20–26, 2001, 4258, 99.

8

Nanoractors in Stem Cell Research

Volker Mailänder and Hubert Schrezenmeier

The focus of this chapter will be twofold: First there is the aspect of stem cells as nanoreactors and how regenerative medicine can explore and harness these metabolic miracles for repairing organ functions and tissue loss. As an introduction for the reader who is not involved in stem cell research, we will describe how the stem cell concept evolved from basic experiments in the hematopoetic system. Based on this preclinical work hematopoetic stem cell transplantation was developed as a therapeutic option for many previously untreatable diseases. We will then describe what led to the discovery of new stem and progenitor cell types during the last decade and how they might be used for repairing impaired functions of organs and tissues. Here we will focus on a highly interesting stem cell population, mesenchymal stem cells (MSC), also called mesenchymal stroma cells.

Secondly, we will explore the terms for interaction of nanoparticles as they get in contact with mammalian stem cells and how they are internalized into stem cells and other differentiated cells. Nanoparticles can be used for various applications: selection of stem cells, detection of homing and trafficking of stem cells by labeling them with nanoparticles, as therapeutic agents that alter the differentiation potential of stem cells and aid in enhancing the therapeutic options in stem cell therapy. Nanoparticles can influence cellular functions and can be used as carriers for DNA and drugs. Prerequisites on both sides—the nanoparticles and the (stem) cell biology side—will be discussed.

Section 8.1 to 8.4 is a primer for the reader interested but not daily involved in stem cell research, in order to understand the discussions going on in this field. Readers only interested in nanotechnology may skip these sections.

8.1 Stem Cells Are a Crucial Cell Population in Animal and Human Organisms

In higher organisms like in animals and humans that reproduce by fusion of two germ cells an omnipotent cell is formed (for definition of potency see Table 8.1). This cell will divide and the offspring of the fertilized ovum will give rise to cells which are toti- or omnipotent (i.e., that can form any type of tissue including germ cells and these will form a new organism). The most exciting example is exemplified by monozygotic twins where it happened by chance that two cells of the first daughter generation of the fertilized ovum were divided spatially and developed into two complete beings. The omnipotency of germ cell tissue is also seen in malignant tumors—teratocarcinomas—that can form every type of tissue and a mixture of disorganized tissues is found in these tumors. These include hair, teeth, skin, but also liver, kidney, and heart muscle tissue. Cells derived from the earliest stages of embryogenesis have been investigated for their potential in regenerative therapy [1]. But the use of these embryonic stem cells remains controversial. In animal studies these cells have shown that they can form tumors in mice after transplantation (e.g., after implantation into an area of myocardial infarction) [2].

Pluripotency is not always clearly separated from toti- or omnipotency by many authors. It can be defined as the ability to give rise to all types of cells (including germ cells) but having lost the ability to form a new organism. Multipotency is the ability to give rise to different types of committed cells (also called differentiated cells), but not to germ cells. The differentiation potential of these cells is limited to a defined set of cell types. The best known and studied example is the hematopoetic stem cell that gives rise to diverse cells that exert very different functions in our body, like transportation of oxygen (erythrocytes), defense against bacteria, viruses, and nonself cells with a broad range of specialized cells (neutrophils, macrophages, lymphocytes, dendritic cells, and so on) and cells that aid in stopping bleeding (thrombocytes) [3, 4].

Table 8.1
Definitions of Terms Used In Stem Cell Biology

Term	Definition
Self-renewal	Ability of a cell population to give rise to the same undifferentiated cell type for an extended time (e.g., lifelong).
Consecutive transplantability	Cells harvested from a donor animal (D) can repopulate and reconstitute the cells in a recipient animal (R1) after the donor animal has been deprived of the investigated cell population. After a relatively long time in the recipient (R1) the cell type can be harvested again from the first recipient (R1) and transferred to a second recipient (R2). These cells of donor origin (D) give also rise to a self-sustaining tissue for a prolonged time in the R2.
Asymmetric cell division	After cell division the two cells have different abilities: one has all the properties of the mother cell while the other cell has started to differentiate into another cell type and does not display the stem cell features of self-renewal and consecutive transplantability.
Omnipotency/ Totipotency	Giving rise to a variety of differentiated cells and also germ cells, having the ability to form a new organism as a whole, and even extraembryonic tissue (fetal placenta).
Pluripotency	Giving rise to a variety of differentiated cells and also germ cells but without the ability to form a new organism as a whole.
Multipotency	Giving rise to a variety of differentiated cell excluding germ cells. The differentiation potential is restricted to some cell types, usually within the same germ layer (ectoderm, mesoderm, entoderm).
Undifferentiated cell	Cell type which main task is to give rise to differentiated cells. Although sometimes it performs specific metabolic functions, or other functions not seen in the progenitor or offspring in an animal, this is not the main purpose of this cell population.
Differentiated cell	Cell type which performs a specific metabolic or other function in the context of a tissue or animal.

Stem cells are necessary for humans and animals to replace cells that have ceased because they have reached the end of their life span or because some kind of injury occurred. The life span of the organism is significantly higher than that of many of its differentiated cells. This is easily illustrated by the relatively short life span of neutrophil granulocytes which die after a few days or thrombocytes that have a life span of about 10 days, with erythrocytes having a life span of 120 days while a stem cells persist lifelong. This means, for example, that in an adult person the content of erythrocytes of about 40 to 50 ml of whole blood (about 3×10^9 erythrocytes per kg of body weight) has to be produced every day by the hematopoetic stem cells. Therefore, roughly 7g hemoglobin has to be synthesized every day just in order to compensate for the loss of erythrocytes that were removed as they had aged. After a severe bleeding event the erythropoesis can even produce up to ten times of this amount.

In the following sections we will only consider adult stem cells as everybody is carrying around a large quantity of these cells and some of them have been used successfully in clinical applications. In addition to this physiological replacement of aged cells, stem cells are in particularly necessary for repair of tissues after various kinds of injury.

8.2 (Stem) Cells as Nanoreactors

Stem cells and other cells are objects of microscale dimension; typically they are several micrometers in size. Nevertheless, many functions are in the nanoscale range. Cells themselves comprise many subcellular compartments which exert specialized functions for the cell itself and for other cells. These are best illustrated by electron microscopy. Even the simplest looking compartments that are surrounded by a lipid bilayer fulfill important functions in an orchestrated way inside the cell. One of these more simple compartments is, for example, a lysosome in which material that is taken up from the outside of the cells is exposed to a low pH [5]. This is achieved by H+ pumps that are integrated into the lipid bilayer. But for trafficking of these nanocompartments, and for selection of the correct material that should end up in such lysosomes a miraculous machinery is needed [6]. Other delicately regulated sets of protein are involved in uptake or transport of specific molecules (e.g., transferrin and ferroportin for the transport of iron) [7]. The uptake of such transport proteins or other material from outside the cell involves a whole set of molecules: a receptor that binds the transport molecule, an intracellular domain of this receptor that signals that the transport molecule has bound to it, several proteins that bind to the lipid bilayer from inside the cell (the cytosolic side) and form a grid (e.g., clathrin) that bends the lipid bilayer. By bending the lipid bilayer more and more a sack-like structure is formed. Finally, an energy-driven process is needed to clip the sack and close the mouth of the sack. This is achieved by dynamin [8]. In other nanoscale compartments—like the endoplasmatic reticulum—new proteins are synthesized by assembling amino acids. These are then packed into vesicles transported through a stack of lipid bilayers—the Golgi apparatus—finally budding from there. They are transported to the cell membrane where they are exocytosed [9]. In this way larger molecules like collagen can be synthesized in the cell and can then be released to the extracellular space where these proteins can aggregate into fibers for bone formation or ligaments [10]. Smaller molecules are sometimes synthesized in the cytosol—the noncompartimentalized rest of the cell. Some of these products can diffuse through the lipid bilayer or specialized membrane proteins form pores. These pores are truly nanoscale entities by themselves as they are often comprised of several subunits that form the pores, and the size and chemicophysical properties of the pores determines the type of molecule that can pass through [11].

8.3 The Concept of Stem Cells is Born: Definition of the Hematopoetic Stem Cell

All organisms live in a steady-state equilibrium of losing or dying cells and replaced or divided cells. A first definition of the term "stem cell" can be given as follows: The main function of stem cells is to be present in an organism in such quantities so that they can give rise to sufficient numbers of various tissue cell types. The function of these differentiated cells is inherited by the genetic and epigenetic setting of the stem cells and therefore stem cells are the key cell type in order to understand and influence the functions of these differentiated cells. Stem cells themselves are not primarily thought of as metabolic entities. Although they are not primarily addressed as bionanoreactors their descendants' key function is, in many cases, a specific metabolic task. In order to understand better how the concept of the "stem cell" was defined we will recall the crucial experiments to its definition.

For many years the best defined stem cell was the hematopoetic stem cell which gives rise to all types of blood cells [e.g., red blood cells, monocytes/macrophages, lymphocytes, granulocytes (neutrophils, basophils, eosinophils)]. All these cells have a limited half-life which can be very short (e.g., 48 hours for neutrophil granulocytes). In the early 1960s McCulloch and Till had performed the crucial experiments [12]. They had shown that mice will die after a defined dose of irradiation. The causes of death at the lowest lethal dose were mostly infections because the differentiated cells of the immune system disappeared from the organism as they were not replenished anymore. When these mice were given bone marrow from a nonirradiated mouse these mice would survive. All red blood cells, and also the leucocytes, were donor-derived cells. The spleen of these mice was punctuated and they were able to show that "islets" of hematopoiesis had formed and each islet was derived from a single progenitor cell as irradiation had caused specific chromosomal changes. Just by examining these cells, and also by looking at bone marrow aspirate microscopically, it was evident that not all cells would contribute to the donor-derived cell populations. Therefore, bone marrow aspirates as a mixture of cells needed to be divided into subpopulations. Two inventions paved the way to determine the hematopoetic stem cell: First raising antibodies against specific surface antigens. termed cluster of differentiation (CD), made it possible to define subpopulations not merely by differences in cell morphology. Secondly the invention of a machine, fluorescent-activated cell sorter (FACS), that could sort cells that were tagged with a specific fluorescent antibody against such a CD molecule made it possible to sort out a few cells from thousands and millions of nonstem cells [13]. This technology takes advantages of the stem cell population characterized by the presence of markers like CD34, CD133, CD117, and the absence of lineage-specific markers which appear later during differentiation. Thus, each stage of differentiation is reflected by a specific immunophenotypic pattern of presence or absence of surface markers.

However, the litmus test for a cell being a true stem cell is still a functional test: hematopoetic stem cells must demonstrate their ability for long-term reconstitution in a recipient and their ability to be transplanted consecutively. For example the hematopoetic stem cells can be transplanted from a donor mouse to a recipient mouse (first generation, R1) and in the recipient mouse R1 cells of the hematopoesis of the donor D will be found. The donor cells (D) in the bone marrow of recipient R1 can be aspirated again even after a long time and can be transplanted into a second generation of recipients (R2). R2 will not only be rescued from the lethal irradiation but will also bear donor cells (D) lifelong. While there are still many cells in the bone marrow that can give rise to differentiated cells, but will disappear after a few weeks or months, stem cells will replenish not only the pool of differentiated cells but will also keep the numbers of undifferentiated cells constant lifelong (see Figure 8.1). This first feature of stem cells has been termed consecutive transplantability. Thus, the potential of the long-term repopulating HSC (LT-HSC) is illustrated best by experiments in which a single cell can repopulate a lethally irradiated donor for the rest of its lifespan [14, 15]. This is achieved by the hematopoetic stem cells by asymmetric cell division [16]—another important feature of stem cells. Stem cells will divide into two cells with one of these cells still having all the characteristics of a long-term repopulation stem cell (LT-HSC) while the other offspring will be predestined to divide further and this cell and its offspring will lose the undifferentiated stem cell phenotype and will give rise to differentiated cells. Another

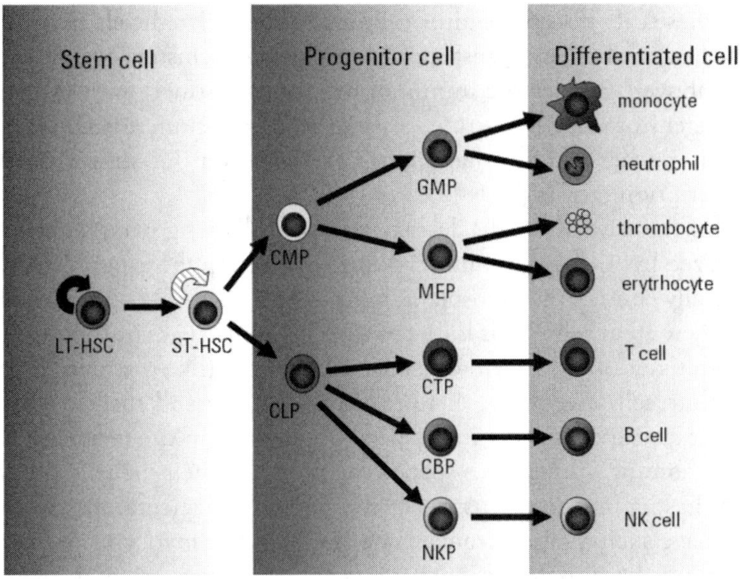

Figure 8.1 Differentiation of hematopoetic stem cells. The long-term repopulating stem cell (LT-HSC) is the most primitive stem cell that can be serially transplanted with a high capacity for self-renewal as depicted by the half-circle arrow. The short-term repopulation stem cell can still produce all cell types but cannot sustain to produce offspring for a prolonged time and therefore cannot be serially transplanted (ST-HSC). The progenitor cell compartment—that does not show asymmetric cell division—is comprised of cells that can only give rise to myeloid (common myeloid progenitor, CMP) or lymphoid (common lymphoid progenitor, CLP) offspring. There is an even higher restriction in the following progenitor cell populations. Differentiated cells comprise the last compartment (T, B, and NK cells). MEP = megkaryotic/erythroid progenitor, GMP = myelomonocytic progenitor. (*After:* [4].)

cell population, the short-term HSC (ST-HSC) (see Figure 8.1) can give rise to a population of differentiated cells for a few weeks. However, this is not a sustained production. After a few cell divisions these cells will disappear and cell counts for differentiated cells will decline in peripheral blood after a while. Another cell type is the committed progenitor cell which implies that it will not be able to replenish their own pool indefinitely (the pool of progenitor cells) by themselves and have only a limited differentiation potential. For example there is a cell type for the myeloid lineage of cells (common myeloid progenitor; CMP), one for the lymphoid lineage (common lymphoid progenitor; CLP), and so on (see Figure 8.1).

All of these different progenitor cells and differentiated cells have been characterized by antibodies against surface markers [see clusters of differentiation (CD) above.] Besides the immunophenotypic distinction of different stem and progenitor cell compartments there has been an increasing interest in expression of other markers that are not presented on the surface and are mostly transcription factors [4, 18].

None of the new stem cells described in the following paragraphs have been scrutinized by such a rigorous procedure paralleling the experiments for HSC. Especially consecutive transplantability (D $\rightarrow$ R1 $\rightarrow$ R2) is not shown for most of these stem cells. This is due to three circumstances: it is not trivial to get enough tissue-specific stem cells from a recipient (like the muscle stem cell, the satellite cell). Secondly, there is mostly no easy and specific way to delete all or at least most of the tissue-specific stem cells in the recipient, and furthermore, homing of the tissue-specific stem cells to their niche is not easily achieved. Lastly, most other tissues as compared to the hematopoietic system do not have such a high turnover rate (e.g., cardiac myocytes, neurons) and therefore it is hard to detect if the stem cells have given rise to differentiated cells. It seems that in the hematopoetic system it was a lucky chance to develop the concept of stem cells as all these obstacles do not apply to HSC.

Nevertheless, many groups have defined tissue-specific stem cells by applying less rigid criteria. These would be lifelong persistence of a specific cell type and the ability to regenerate the tissue or specific cell types of that tissue.

8.4 "New" Stem Cell Types

Until the mid-1990s the concept of stem cell properties was exemplified by hematopoetic stem cells (see above). The regenerative potential of different tissues was categorized into three groups: (1) tissue with a high cell turnover and a nearly lifelong capability of healing tissue damage ("mitotic tissue," e.g., hematopoesis, intestinal epithelia, dermal epithelial cells). (2) Tissue with a low cell turnover but with a limited capability of regeneration of tissue when tissue damage occurred [i.e., these cell losses can be repaired ("facultative mitotic/reversible mitotic tissue," e.g., liver, kidney)]. (3) Tissue without the ability to heal tissue damage [i.e., these cell losses cannot be repaired ("postmitotic tissue," e.g., heart muscle, neurons)]. Stem cells were thought to be present in mitotic tissues only.

In the late 1990s there was accumulating data that cells of a specific tissue type may be able to transdifferentiate into cells of other tissues [19–21].

They would therefore be pluripotent—one of the characteristics of stem cells (see Table 8.1). We want to exemplify this work that has been done in several groups by the work of Mezey et al. When Mezey et al. were trying to elucidate which cells are the progenitors of a specific phagocytosing cell type of the brain, they were transplanting HSC into mice—the prototype of a phagocytosing cell. It is one of the well-known cell types that HSC gives rise to. In these experiments they found that HSC would not only differentiate into these phagocytosing cell types but also in other cells surrounding the neuronal cells in the brain and even into neuronal cells themselves.

The possibility of "turning blood into brain" [22] or "turning brain into blood" [20] turned over the paradigm of postmitotic tissue and opened the possibility that damaged neurons as one of the hardest targets for tissue regeneration could one day be replaced by cells of quite another origin—hematopoetic stem cells. This phenomenon was termed transdifferentiation. The publication of this work resulted in an avalanche of work on transdifferentiation of different cell types into a variety of terminally differentiated tissue cells of another tissue type, that is in patients that had undergone an allogeneic HSC transplantation for a hematopoetic disease (mostly some type of leukaemia) male donor cells with a Y chromosome were found in the heart muscle of female recipients therefore proving that male donor-derived cells had integrated themselves into heart tissue [23]. Donor cells were also found in liver tissue, intestinal epithelium, and other tissues [24]. More and more cell types were described that were able to differentiate or transdifferentiate into specific tissue cell types (for an overview see [25]). However, the principle of transdifferentiation has been challenged later on since several alternative mechanisms could explain the experimental findings [26, 27]. The way these cells get integrated into the target tissue remains an issue of debate as some groups have shown that not transdifferentiation, but fusion of a dying (recipient) tissue cell (e.g., liver cell) with a (donor) progenitor cell occurs, and results in integration of donor cells into recipient tissues [28, 29]. Thus, "cell fusion causes confusion" [30].

Many of these progenitor cells have been termed "stem cells" by the authors although they do not fulfill all the criteria that apply to stem cells like hematopoetic stem cells do. Most of them show an interesting variety of cell types into which they can be differentiated in vitro or in vivo. Many of them are also highly proliferative. On the other hand, data on asymmetric cell division [i.e., that there are still stem cells after induction of (trans-) differentiation] is mostly lacking. Also, consecutive transplantability has not been shown. One of the main drawbacks of many of these "stem" cell populations is that they cannot be accessed without destruction of tissue and that the quantity of cells harvested by these procedures is rather low.

Therefore we want to focus on a well-accessible progenitor cell population that is also highly interesting as it can be expanded or cultured in high cell numbers more easily than others: mesenchymal stem cells (also called mesenchymal stromal cells thereby avoiding the term stem cell [31]).

8.4.1 Mesenchymal Stem Cells (MSC)

Friedenstein et al. introduced an ex vivo assay for examining the clonogenic potential of multipotent marrow cells [32]. One of the cell types specified in this assay were stromal cells, referred to as colony-forming unit-fibroblasts (CFU-F). In 1999 these cells were shown to be able to differentiate into a variety of stromal tissue types, namely adipose tissue, cartilage, and bone [33], and have been renamed as mesenchymal stem cells by these authors. Furthermore differentiation into neuronal cells [34] or cardiac muscle has been shown (see Figure 8.2) [35]. They can be easily harvested from bone marrow and many other tissues by plastic adhesion and cultured in minimal media. The precursor frequency is estimated to be in the range of one MSC precursor per 10^5 to 10^6 bone marrow cells. Morphologically, these cells show a high degree of pleomorphy. Immunophenotypically, these cells are characterized by a set of markers (expression of CD9, CD73, CD90, CD105, CD146, CD166 while lacking CD34 and hematopoietic lineage-specific markers). Their high-proliferation rates gives yields of about 1×10^8 cells after a three to four week expansion from a single bone marrow aspiration [36]. Even after this prolonged time under cell culture conditions MSCs keep their multipotent differentiation potential. Therefore, they have gained a lot of interest during the last few years.

Trials with MSC for tissue repair have been conducted. In these trials the differentiated cells restore normal organ function (like bone formation for osteoblastic differentiation [37], cartilage after chondrogenic differentiation), and the cells are used as potent producers of extracellular matrix which are mostly different types of collagen and other extracellular biopolymers. In these trials patients with a defect of bone formation called osteogenesis imperfecta had less bone fractures when they were treated with MSC. They have also been used experimentally for cardiovascular repair [38, 39], treatment of lung fibrosis [40], spinal cord injury [41], and recellularization of heart valves [42].

MSC have also been investigated for support of hematopoetic engraftment after peripheral blood HSC transplantation [43]. It was

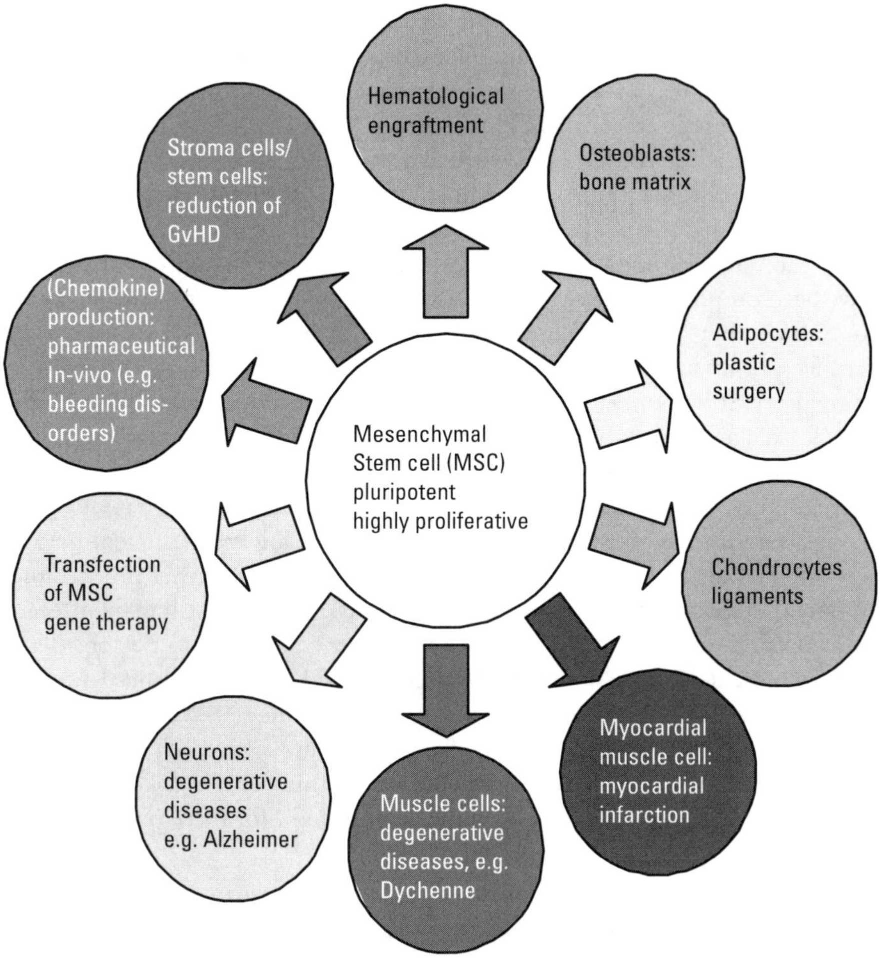

Figure 8.2 Illustration of the differentiation potential of mesenchymal stem cells and their possible applications for regenerative medicine.

speculated that MSCs produce a hitherto unidentified humoral factor that is boosting the engraftment and the proliferation of HSC.

After it was discovered that MSCs were not rejected even in allogeneic settings, their immunosuppressive potential was elucidated in vitro and in animal studies. Therefore studies on suppression of allo- or autoimmune, diseases and other conditions in which immune cells play a predominant role in the pathogenesis have been conducted [44]. Treatment of graft-versus-host disease—a situation in which the transplanted immune cells from the donor recognize recipient cells as "foreign" cells and try to destroy them—has been done in humans. These trials are among the forefront of cellular therapies [45]. In a study of the European Group for Blood and Marrow Transplantation (EBMT) evaluating cotransplantation of ex vivo expanded MSC in conjunction with transplantation of HSC there were less acute and chronic graft-versus-host disease, thereby preventing this complication [45, 46].

Most interestingly in the context of stem cells as nanoreactors, MSCs have been proposed as potential precursors of tumor stroma and as MSC home to tumors they could be used as vehicles for tumor therapy [47]. Here transfection of suicide genes would transform MSC into a Trojan horses by using the MSC as a metabolic production site of noxious agents. Other studies have looked at the possibilities of MSC getting trapped in lung capillaries [48]. There has also been investigations of migration of MSC (e.g., in the brain) where they could produce specific molecules (e.g., neurotransmitters) [49].

Overall, there is enthusiasm for the use of adult stem/progenitor cells for tissue repair in end-stage diseases for which there are either no therapies or only therapies with rather insufficient results. However, still many precautions on potential adverse effects of stem cell therapy have to be considered [50]. As research on tissue repair by stem cells go on we will need tools to modulate functional properties of stem cells and tools to track stem cell fate, and we will need better defined and purely selected stem cell populations. Nanotechnology could help to provide all of this. In the following paragraphs we will discuss how nanoparticles can be used for these purposes.

8.5 Nanoreactors/Nanoparticles and Mammalian (Stem) Cells

In the following sections we will explore how nanoreactors and nanoparticles have been used in conjunction with stem cells and differentiated mammalian cells, what is being developed, and which nanodevices could be imagined in the future.

8.5.1 Prerequisites for Polymers and Other Components of Nanoparticles and Nanoreactors for Use in Stem Cell Biology

The use of nanoparticles and nanoreactors in mammalian (stem) cell research must take into account that the nanoparticles with all their components should either not alter the specific functions of the target cells as in diagnostic use, or should alter the biological functions only in a defined way as in theraputic use. Whenever nanomaterials are in contact with cells—not only stem cells but also differentiated cells—they should be nontoxic (unless they are used as chemotherapeutic agents in tumor therapy) and should not affect the growth kinetics of the cells. The cells should still perform their regular functions, such as synthesis of extracellular matrix, production and release of cytokines, metabolic functions like degradation of products, signaling to neighboring cells, tissue repair, and so forth). These investigations are only infrequently carried out [51]. These functions will have to be determined depending on the type of cell investigated. Furthermore the ability of the stem cells to migrate and interact with other cells (e.g., by surface expressed receptors or by paracrine hormone secretion) should not be altered in an undesired way [52]. In stem cell biology further functions should not be distorted including the ability of self-renewal—yielding offspring with the capacity to give rise to the same variety of differentiated cells as the mother cell did—and showing asymmetric cell division [53]. Furthermore, the variety of differentiation potential is to be assessed. Even when undifferentiated stem cells after exposure to nanomaterials are still present after several passages in vivo or in vitro there is still the possibility that the differentiation potential of the stem cells is altered in an unintentional way by the influence of the nanomaterial. There is a debate about whether mesenchymal stem cells (MSC) lose their chondrogenic differentiation potential after incubation with superparamagnetic iron oxide particles formulated with dextran (SPIO [54, 55]). Interestingly, this has also been demonstrated for quantum dots which are made of cadmium and selen [56] thereby pointing towards the possibility that this effect on chondrogenesis of MSC may not depend on the type of reporter molecule or assembly in the nanoparticles but other components or properties may be effective. On the other hand, these findings open the possibility to potentially steer the differentiation of MSC away from the chondrogenic path if this is not the desired differentiated cell population.

8.5.2 Components of Nanodevices to Be Considered in Affecting (Stem) Cell Functions

Whenever an influence of nanoparticles on cell behavior and especially on stem cell behavior is demonstrated, the causative component of the nanodevice should be evaluated. Most nanomaterials are composed of different substances and nanoreactors can be even more complex than the rather simple nanoparticles used in many applications. Even so, these and future nanomaterials will have several of the following components involved: They may be constructed of inorganic material (superparamagnetic iron oxide in

SPIOs, CdSe for quantum dots, mesoporous silica, gold, and so forth) mostly used as a reporter substance, an organic (shell) material (e.g., dextran, poly-butylcyanacrylate, poly-L-lactide, poly-ε-aminocaprolactone, polystyrene just to name a few, see Section 8.5.4, "Polymers Used for Applications in Mammalian Cells"), and surface modifications (like amino or carboxyl groups of comonomers—therefore chemically bounded into the polymeric shell material, see Figure 8.3, or absorbed to the surface by layer-by-layer technique). All of these will need to be evaluated in terms of their influence on (stem) cell functions (e.g., the monomer used for synthesis of the nanoparticles) [57]. Also further modifications like amino acids, peptides,

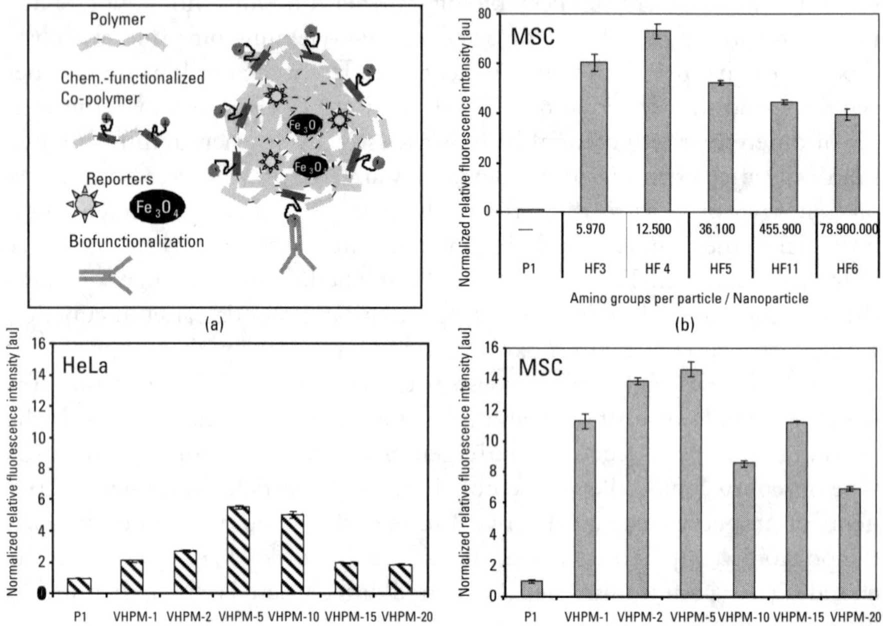

Figure 8.3 (a) Components and assembly of polymeric nanoparticles as obtained by the miniemulsion process. (b) Uptake of amino-functionalized nanoparticles into mesenchymal stem cells illustrating that amino-functionalized nanoparticles are more effectively taken up but an increase in amino groups is not promoting cell uptake. (c, d) comparison of HeLa (carcinoma cell line) and MSC in terms of uptake of carboxyl functionalized nanoparticles. Numbers at the end of the particles' name denote the amount of comonomer (acrylic acid) used. MSC showing a higher uptake than HeLa cells. Comparison of (d) with (b) illustrates that amino functionalization is superior in terms of nanoparticle uptake. P1: unfunctionalized particle for comparison. For details, see [158].

proteins or ligands will alter the biological behavior of the nanomaterial [58] by selectively being absorbed to specific cell types or even internalized by these cells. These later modifications will mostly modulate biological responses intentionally. Also other components like surfactants, not-reacted monomers, stabilizers (like citric acid), or other components involved in the manufacturing process (e.g., traces of the continuous phase like cyclohexane) may influence (stem) cell functions. We have shown, for example, that the length of the polymer chain of polybutylcyanacrylate is important for nanoparticle toxicity [59].

To complicate things even more, it may be of importance for the biological function if the component is located on the particle surface or if the material is encapsulated inside of the nanoreactor and if the material is accessible by a leaky shell. Substances encapsulated in a nanomaterial that is not accessible from the outside and as long as the encapsulating material is nondegradable under physiological conditions, the encapsulated substance should not exert any biological effects.

8.5.3 Synthesis of Nanoreactors and Nanoparticles for Use in (Stem) Cell Biology and Therapy

For the synthesis of nanoreactors and nanoparticles several synthetic processes have been utilized. The method of choice for the synthesis of nanoparticles and nanoreactors in our group is the miniemulsion process which is highly adaptable to a variety of monomers, surface functionalizations, and materials incorporated (see Chapter 2).

In order to analyze the fate of the nanoparticles in biological environments, it would be desirable to encapsulate a fluorescent reporter. However, due to diffusion processes during the standard method of emulsion polymerization, these methods do not easily allow for the encapsulation of reporter molecules like fluorescent dyes or other substances that are to be encapsulated for a later release. For the preparation of fluorescent nanoparticles and also for the encapsulation of various (bioactive) substances, it was shown that the miniemulsion process is excellently suited [60]. The monomer, a dye, a hydrophobic agent for the suppression of Ostwald ripening and possibly other substances, which are to be encapsulated, are dispersed by applying high shear. Then, a stable mixture of small, stable, and narrowly distributed nanodroplets in an aqueous surfactant solution is formed. The nanodroplets can be perceived as nanoreactors in which the monomer is polymerized afterwards, leading to particles with a narrow-size distribution in the range of 50 to 500 nm in diameter [61].

8.5.4 Polymers and Surface Modifications Used for Applications in Mammalian Cells and Medical Applications

Different polymers have been investigated for their use in targeted nanoparticulate drug delivery. This application is one of the most promising techniques for increasing the efficiency of drugs [62]. With the incorporation or adsorption of drugs into a carrier system, the drug can be effectively protected from degradation or metabolization after its administration so the drug can be released slowly and gradually [63, 64]. As carrier systems, different types of polymeric nanoparticles have been proposed. For some applications, like vaccination, particles of nonbiodegradable polymers like poly(methyl methacrylate) (PMMA) are proposed. For sustained-release application, (bio)degradable polymers like poly(alkyl cyanaoacrylate)s (PACA) or polyesters (e.g., poly(ε-caprolactone), poly(D,L-lactide)) are the first choice [64–68]. Despite the protection of the drug, not all of it reaches the site where it is required. Several reasons can interfere with the effective delivery of the drug to the desired destination. The drug may lose its protection due to a rapid degradation process of the carrier system, the carrier system shows unspecific affinity to any type of tissue, the particles might be opsonized by absorption of serum proteins and phagocytosed by macrophages or they may be excreted from the organism.

Targeting to or avoidance of specific cell types—mostly macrophages—on one hand and simultaneous protection of the particles on the other hand should overcome these obstacles. The most common approach to protect particles from opsonization is the formation of a hydrophilic corona, for example from PEG (see Figure 8.4) or carbohydrates, around the particles [69–72]. This shell minimizes unspecific adsorption of proteins on the particle surface, which renders the particles "invisible" for macrophages ("stealth" effect). Therefore the particles remain unaffected in the circulation, thereby increasing the chance that the site of disease is reached by the particles, and as metastases of a tumor show an enhanced permeability with a decreased rate of clearance, more nanoparticles are trapped at the site of the disease ("passive targeting") [73–75].

8.5.5 Selection of Stem Cells for Transplantation

There are three main sources for hematopoetic stem cells: first the bone marrow, secondly they can be mobilized into the blood stream after chemotherapy or/and after treatment with a growth factor (e.g., G-CSF or GM-CSF), or they can be acquired from neonatal blood (cord blood cells) [76]. In all these cases not only stem cells but also other cells like immune cells (T cells, B cells,

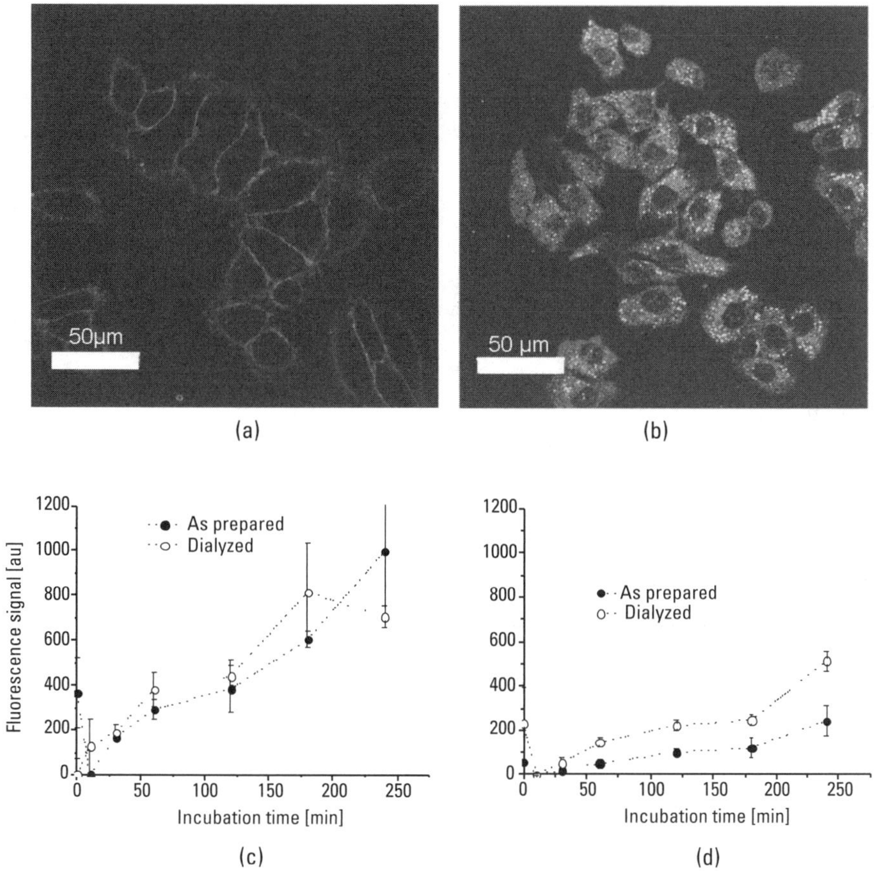

Figure 8.4 Making NP invisible (stealth): (a) Cell uptake of polybutylcyanacrylate (PBCA) nanoparticles; (b) Uptake of PBCA nanoparticles as detected by laser-scanning microscopy comparing a negative control (no nanoparticles) with unfunctionalized nanoparticles; (c, d) comparison of unfunctionalized and PEG-functionalized nanoparticles with PEG-functionalized PBCA particles are protected from uptake (for details, see [59]).

monocytes, neutrophils), red blood cells, and in the case of bone marrow also stromal cells are collected. In fact, true HSC only comprise a minority of all white blood cells in these cellular products [76]. Sometimes this may be advantageous. For example, donor T-cells help to eliminate recipient hemopoietic cells and therefore create space for engraftment of donor cells in the bone mesh work. Also donor stromal cells can facilitate engraftment—one of these cell types are mesenchymal stem cells [43]. On the other hand donor T-cells can attack cells of the recipient and can eventually have lethal consequences (graft-versus-host disease, GvHD) [77, 78]. Finally, the stem cell graft might also be contaminated with cells of the underlying disease (e.g., tumor cells or leukaemia cells) in the setting of autologous transplantation. Therefore "purification" of stem cells has been developed. While the earliest methods used negative selection (i.e., depleting the product of the unwanted cell populations by "rosetting"—a technique where T-cells would attach to sheep red blood cells), positive selection for stem cell markers like CD34 or CD133 was applied later on [79]. These later procedures were done by using (superpara-)magnetic micro- and nanoparticles. The selecting antibodies were attached to the nanoparticles, these bind specifically to their antigen on the cell surface and the marked cells are trapped in a magnetic field on an iron wool column [80, 81]. By removing the magnetic field the cells can be recovered from the column. This technique has enabled selection of stem cell grafts with a very high purity of CD34+ or CD133+ cells and is one of the earliest examples of nanotechnology that has now been employed for more than a decade as a routine procedure in hematopoietic stem cell transplantion.

In some situations an ideal matching stem cell donor cannot be found and therefore even donors are considered who are not (fully) histocompatible with the recipient. A HSC preparation containing a substantial number of immunocompetent T-cells would inevitably result in a severe, life-threatening GvHD. Immunomagnetic selection allows preparation of a CD34+ cell preparation almost completely depleted of contaminating T-cells. This allows transplantation even in situations with only a haploiddentical match (e.g., transplants with donation of stem cells of a parent for the child) [82]. While in these cases completely depleting the immune cells is the only way to transplant, in other circumstances certain types of immune cells are desirable (NK-cells, dendritic cells). They aid in defending the host from infections [83–85]. By taking out only the most aggressive subtypes of cells (mostly B and T-cells) these grafts can be "tuned" as desired [86]. This will aid in minimizing the side effects of these therapies. On the other hand selection of stem cells has been propagated for depleting contaminating tumor cells from the stem cell grafts [87]. Although this seems to be a desirable approach, selection of stem cells is not performed in all transplants and the risk reduction is discussed controversially [87–89].

8.5.6 Diagnostic Use of Nanotechnology in Stem Cell Biology

There are many new tools based on nanoreactors and nanoparticles that have been reported in the literature. In this section we want to focus on two applications where a contact between stem cells and other mammalian cells and nanoparticles/nanoreactors have been established by research and development efforts during the last decade: nanoparticles as contrast agents for MRI and nanocapsules or droplets as nanoreactors.

8.5.6.1 Nanoparticles and Nanocapsules as Markers for MRI

Cell tracking and homing studies are crucial for understanding the fate of new stem cell therapeutics. Understanding these mechanisms will augment the benefit of these novel therapeutic approaches by tailoring the cells and their migratory potential for a specific indication, for example migration to an infarcted tissue in the heart. In order to detect homing and migration of transplanted cells, techniques like bioluminescence [90, 91], radioactive substrates [92], near-infrared fluorescence [93, 94], and labeling with magnetic resonance imaging (MRI) contrast agents are applied in small animal studies. Of these, only labeling with radioactive agents and MRI contrast agents are suitable for studies in humans as well.

For MRI mainly two classes of substances are used as contrast agents. While gadolinium can be used for cell labeling and detection by MRI [95] most approaches utilize superparamagnetic iron oxide particles (SPIO) for this purpose. SPIOs are negative contrast agents and are commercially available and FDA-approved for use in humans [96, 97]. However in order to achieve intracellular uptake of nanoparticles, transfection agents are needed in most studies [96, 98–101]. These transfection agents like Superfect, DOTAP, Lipofectamin [98, 100], poly-l-lysine (PLL) [100, 101], or protamine [99] are mostly cationic, positively charged molecules. Higher concentrations of these agents are toxic [102, 103] and—with the exception of protamine—are not approved for clinical use. Avoiding a transfection agent in the process of cell labeling seems to be favorable as this would simplify the approval of such studies in humans.

Interestingly, it has been shown that Feridex-labeled cells can be followed in the living animal for up to three weeks [97]. In Resovist (generic name: ferucarbotran) magnetite nanoparticle aggregates are stabilized by carboxydextran, in Feridex (generic name: ferumoxides) by nonfunctionalized dextran. Triggered by observations in light microscopy studies after Prussian blue staining we have shown that Resovist was taken up spontaneously, (i.e., without the need for a transfection agent), and that the load of intracellular iron can be as high as with the use of a transfection agent.

As cellular targets we chose human mesenchymal stem cells (MSC) and a tumor cell line [104–106]. Focusing on the differential behavior concerning the uptake into cells between the two nanoparticles Resovist and Feridex, we hypothesized that the carboxyl groups in the carboxydextran of Resovist are responsible for a more efficient uptake. By using the functional monomer acrylic acid as a comonomer, different densities of carboxyl groups on the surface of the polymeric nanoparticles can be obtained by varying the amount of the comonomer in the miniemulsions (see Figure 8.3, for details see [107]). Hereby experimental polymeric nanoparticles with increasing amounts of carboxyl groups on their surface as model systems can be synthesized, and the uptake behavior of nanoparticles can be studied.

For gadolinium—a positive MRI contrast agent—encapsulation is favorable in order to deliver a high concentration of molecules at one site. In the literature, three main different approaches have been followed. Aime et al. [108] presented an agent based on compartmentalization of Gd complexes in apoferritin cavities. Although this approach results in a very high relaxivity of about 80 $mM^{-1}s^{-1}$, the very limited payload per capsule limits the applicability of the technique as a target contrast agent. A very promising approach has been reported by Morawski et al. [109], who incorporated a large number of Gd-DTPA complexes on the surface of perfluorcarbon nanocapsules. They reported rather high relaxivities value of 17 mM-1s-1 and high payloads of up to 100,000 Gd-DTPA complexes per capsule. Reynolds et al. [110] reported a polymeric core-shell nanocapsule approach for achieving high paramagnetic payload but the T1 relaxivity value is not reported. The challenge here is that the shell of these nanocontainers has to be permeable to the water molecules as they need direct interaction with the gadolinium [111], but the gadolinium complex itself should not diffuse through the capsule shell.

We have demonstrated that by the indirect miniemulsion technique and subsequent transferring these nanocapsules to the water phase that polyurethane, polyurea, and crosslinked dextran nanocapsules containing a contrast agent (Magnevist, Gadovist) can be synthesized [112]. The shell of these nanocapsules is highly permeable for water allowing almost free exchange with the bulk water. In comparison with the nonencapsulated agent, T_1 relaxivity measurements revealed a slight decrease of relaxivity in water as well as in human blood after encapsulation of Gd-DTPA (Magnevist) [112]. Since for targeted contrast agents, the limiting factor is the number of local receptors sites, the local relaxivity is governed by the number of bound capsules. Considering the possible high Gd load of a single capsule between roughly 10^{-17} mMol (50 nm) and 10^{-14} mMol (300 nm) while at least maintaining the relaxivity of the embedded agent, the possible enhancement per binding site is significantly boosted compared to conventional paramagnetic extracellular contrast agents. The suggested approach appears not to be limited to the encapsulation of Magnevist as proven by the Gadovist example, and can likely act as a new basis for versatile contrast agents in MRI. A functionalization of the nanocapsules will allow the specific targeting of certain molecular targets for detection of specific diseases. For example, folic acid [113] and transferrin [114] has been used for imaging of tumors overexpressing the receptors for these molecules.

8.5.6.2 PCR in Nanoreactors

Polymerase chain reaction (PCR) is a common method that is used to create copies of a specific region of a deoxyribonucleic acid (DNA) sequence, in order to produce high enough quantities of DNA for further biochemical analysis [115]. A few DNA molecules, which act as templates, are rapidly amplified by PCR into many billions of molecules. In principle, one single DNA molecule is the minimum amount that is needed to perform a PCR experiment. For example, Walsh and coworkers [116] developed a single-tube "hanging droplet" nested reverse transcription PCR. Other research groups [117–120] described the reaction with a single molecule DNA in water-in-oil droplets. The droplets were formed by stirring the reaction mixture with the magnetic stirrer. The sizes of final droplets were polydisperse ranging from 2 to 15 μm. Due to the large size distribution, the number of the initial template differs significantly among these droplets. This leads to the wide variation in the amount of the final product per droplet volume. Therefore, the use of more homogeneously distributed droplets is of high interest, so that only one DNA molecule will be present and a single reaction per droplet will take place without interacting or mixing with other DNA molecules. It is also of high interest to reduce the size of the droplets in order to have only a small volume per one single molecule because then more nanoreactors can be present in the system.

We have shown that a single-molecule PCR reaction can be performed in aqueous nanodroplets as small compartments [121]. This will enable to study the process of PCR in a compartmentalized system. But also accumulation of several hundreds and thousands of copies from a single DNA molecule in an ultrasmall compartment is a highly attractive feature of this approach. This could be used either as diagnostic tool or for transfection in cells as described below. For example it could be imagined that single-strand DNA fragments amplified in such a way together with a cromophor and quencher can be used as a diagnostic nanodevice (see also molecular beacons [122]) in order to detect target sequences in cells with the molecular beacon always encapsulated or bound to the inner surface of a nanocapsule. The advantage of this approach would be that the molecular beacon could be protected from enzymatic degradation and could not diffuse through the entire cell but would be concentrated in a compartment yielding a stronger localized signal.

8.5.7 Therapeutic Options of Nanoreactors and Nanoparticles in Stem Cell Transplantion

Although the hype for gene transfection and gene modification in humans has flattened, the need for correction of genetic defects is still a challenge. Especially in the hematopoetic system like in severe combined immunodeficiency syndrome (SCID) and other diseases like genetically caused bleeding disorders in hemophilic patients. Correction of the underlying genetic defect is still investigated [123]. HSC with the genetic error can be transfected even in vitro, and then reinfused in an autologous recipient thereby circumventing many of the obstacles like targeting the transfection vehicle to the right cell type. Genetic correction with modified viruses have shown to have severe side effects because of the integration of the corrected genes in the chromosomes [124] and by the transduction vehicles–the genetically engineered viruses–themselves. While the integration of the corrected gene into the chromosome is a prerequisite of viral gene therapy, in order to have stable expression of the corrected gene product, side effects caused by the viruses themselves can be avoided. Therefore, nonviral transfection methods are an interesting topic. Here nanoparticles which were loaded with premanufactured DNA snippets or nanocapsules in which the corrected DNA snippet has been synthesized in the way described above can be utilized. While calcium phosphate particles have evolved from the calcium phosphate precipitation method [125], the addition of silica particles to a transfection agent/DNA complex can enhance the transfection efficiency by altering the intracellular localization of the transfected DNA [126, 127]. Especially the escape from the endosome and hereby avoiding lysosomal degradation of the DNA has been addressed by adding polyethylenamine to the nanoparticles or by incorporating photosensitizing molecules like phthalocyanine that disrupt the endosome once illuminated [128]. Also, the combination of gold nanoparticles and electroporation has been used to transfect cells [129]. Although electroporation is only feasible in vitro and cannot be applied in vivo it is highly attractive as high-transfection efficiencies were shown for hematopoetic stem cells–a cell population hard to transfect otherwise [130]. Also, magnet-assisted transfection has been proposed to enhance the uptake of nanoparticles into cells [131]. Furthermore the transfection rate can be varied by nanoparticles made from thermoresponsive material [132].

8.5.8 Enhancing Effectiveness of Nanoparticles and Nanoreactors in Human (Stem) Cells–Understanding and Influencing the Uptake of Nanostructured Materials in (Stem) Cells

Uptake of nanoreactors and nanoparticles into cells is a crucial step for several biomedical applications like cell labeling for magnetic resonance tomography [133–136], transfection (see above), or drug targeting to diseased tissue [137–139]. For the uptake of these nanostructured materials factors influencing two important processes will be discussed in detail here: first there is the interaction to the cell surface (adsorption/adhesion), and secondly the internalization by endocytosis into the cell. Further mechanisms that are involved are intracellular trafficking and release of the materials from endosomes and lysosomes. In order to be able to improve the cellular uptake into cells for advancing the possible biomedical applications it is essential to characterize both the interactions of nanoparticles with the target cells, and the endocytosis of the nanoparticles. The interaction to the cell and the uptake process of polymeric nanoparticles into cells can be influenced by several parameters, such as: a) an amphiphilic polymer which is physically adsorbed on the nanoparticles' surface; b) surface groups which are covalently bound to the nanoparticles' surface; and c) the structure and morphology of the polymer of the nanoparticles themselves. Especially particle size seems to be an important factor in transfection efficiency [140].

8.5.8.1 Uptake Modified by Surface Coating with Polymers

The uptake of nanostructured materials can be efficiently manipulated by using polymers physically adsorbed on the particles' surface. These polymers are mostly cationic, positively charged molecules and lead to an increase in cell uptake. Prominent examples are poly(ethylene imine) (PEI), [141] superfect [142], polyfect [96], lipofectamine [96], effectene [96], and poly-l-lysine (PLL) [96, 142]. The disadvantage of physically adsorbed polymers is the possibility of detachment and eventual toxicity at higher concentrations.

Cell adhesion of the nanoparticles is dependent on the interaction of the nanoparticles with plasma proteins. Coatings of the nanoparticles that minimize the interactions with the plasma proteins are mostly derivatives of dextran [142], but the effect is even more pronounced with polyethylene glycol (PEG) [143], poloxamers, and poloxamines [144–146].

8.5.8.2 Uptake Modified by Covalently Bound Functionalization of the Surfaces

Another possibility of varying the uptake behavior is the functionalization of the particles' surface by using groups which are covalently bound to the particles. This can be either unspecific surface modifications like altering the surface charge (see Figures 8.3 and 8.5), or attaching antibodies or ligands for cellular receptors. Altered uptake behavior in nonphagocytotic HeLa, Jurkat, and mesenchymal stem cells was observed after surface modification of polystyrene (PS) nanoparticles with carboxy or amino groups (see Figure 8.3) [147, 148]. By copolymerization of styrene with the hydrophilic acrylic acid, the amount of carboxyl groups on the surface was varied. A further increase of uptake can be accomplished by transfection agents like poly-L-lysine or other positively charged polymers (see Figure 8.3). This functionality was also engrafted into the surface of the nanoparticles by covalently coupling the carboxyl groups with lysine. The amount of iron that can be transfected by these lysine-modified nanoparticles was even higher than with the nanoparticles with a transfection agent added, and therefore only physically adsorbed.

While carboxy functionalized PS nanoparticles of a size of 100 nm only show a slight increase of total particle uptake after 24 hours, compared to nonfunctionalized PS particles of the same size, also amino functionalized particles can be obtained by a copolymerization process without the need to employ a coupling reaction after nanoparticle synthesis. These particles show

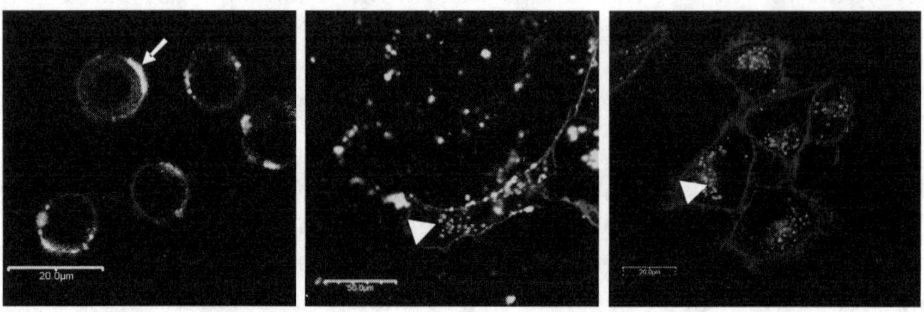

(a) (b) (c)

Figure 8.5 Differences of adhesion and uptake behavior depend on the cell line. (a) KG1a cells (model cell line for CD34+ HSC): amino-functionalized nanoparticles are attached to the cell surface and therefore are visualized in the overlay as yellow spots (arrow). (b) MSC: amino-functionalized nanoparticles are taken up (arrowhead). (c) HeLa (malignant cell line): carboxyl functionalized nanoparticles are taken up (arrowhead). For details, see [159].

an up to 40-fold increased in total uptake. These cationic surfaces resemble the coating of nanoparticles with cationic polymers like SuperFect, and others. Since transfection agents are toxic and not approved for clinical use, and hence applications in human trials and therapeutic interventions are prohibited up to now, covalently bond molecules are desired where the surface of the nanomaterials should be the determining factor for cellular uptake [149, 150]. Amino-functionalized nanoparticles with defined biologically active surfaces were therefore used for an efficient uptake in stem cells without the need for transfection agents [148]. These uptake results demonstrate that nanoparticles with cationic groups on their surface enhanced the uptake rate of particles into cells without using transfection agents compared to uncharged particles into MSCs. An optimal density of amino groups on the surface for uptake into MSC was defined in these experiments in our group.

In MSCs this labeling of the cells resulted in intracellular uptake into compartments resembling endosomes with some small clusters of nanoparticles on the cell membrane. In the other two cell lines–Jurkat (as model for T-lymphocytes) and KG1a (as model for HSC)–there was also a correlation between fluorescence intensity as measured by FACS, and the density of amino groups on the surface. Confocal laser scanning microscopy (see Figure 8.5) and transmission electron microscopy revealed that with these cell lines the fluorescent nanoparticles were attached to the cell membrane in clusters that consisted of nanoparticles, microvilli, and electron-dense material. Thus, extracellular versus intracellular labeling depends on the type of cell used. Labeling of these cell lines can also be achieved by these cationic nanoparticles as the nanoparticle clusters resisted three steps of washing procedures. While attachment of the particles to the cell membrane as the first step seems to be mostly affected by the surface charge of the nanoparticles, differences in the intracellular localization between various cell lines can be explained by different endocytotic/pinocytotic properties of the cell lines HeLa, MSC, KG1a, and Jurkat.

Adhesion of nanoparticles and nanoreactors on the cell membrane and uptake into cells can be mediated also by antibodies, which can be coupled on the nanoparticle surface. Other proteins or shorter peptide sequences like the peptide sequences from Tat peptide have also been used for enhancing cellular uptake [133].

If the unspecific adhesion or uptake of nanoparticles is not desirable, the surface needs to be "shielded." This is done in order to increase the difference between unspecific uptake and targeted uptake. Therefore nanoparticles with poly(ethylene glycol) (PEG) coating has been compared with folic-acid-coated nanoparticles (see also Figure 8.4). This resulted in a

12-fold higher uptake rate for the folic-acid-coated nanoparticles after 4 hour incubation [151]. A higher uptake rate of folate-coated nanoparticles compared to PEG–coated nanoparticles was also found by Oyewumi et al. [152].

Applying the miniemulsion technique, fluorescent dye labeled unfunctionalized and functionalized poly(n-butylcyanoacrylate) nanoparticles were prepared [159]. Amino acid and MethoxyPEG functionalization could be introduced by using aqueous solutions as initiator for the anionic polymerization in heterophase. All prepared particles have sizes smaller than 250 nm and negative ζ-potentials. The molar mass distribution of the polymer is dependent on the acid used as continuous phase and the applied initiator solution. Cells of three lines (HeLa, Jurkat, mesenchymal stem cells) were incubated with the particles. The molar mass of the polymer determines the onset and extent of apoptosis, the total uptake is determined by the size and functionalization of the particles. Different uptake kinetics is obtained with HeLa and Jurkat cells after incubation with the same particle batch. The intracellular particle distribution, visualized by confocal laser-scanning microscopy, does not show significant differences for either of the cell lines or particle batches.

8.5.8.3 Uptake Influenced by the Polymer of the Nanoreactors or Nanoparticle

Hydrophilic and hydrophobic polymeric surface properties of nanoparticles are known to significantly influence cell adhesion in the uptake process. With increasing hydrophobicity of the polymer the attachment on cells and subsequent internalization is enhanced [153]. On the other hand, increasing hydrophobicity is hampering the application of nanoreactors and nanoparticles in cell culture media or in the blood circulation as the tendency of agglomeration increases with increasing hydrophobicity.

Since the polymer itself has an important influence, it is expected that the uptake behavior of nanoparticles can be facilitated if polymers with a high similarity to natural structures are used. The terpene structure is found in many chemical structures in nature, like in essential oils and pheromones. In polymer science it is known as the monomer isoprene. Fluorescent polyisoprene nanoparticles were synthesized by the miniemulsion technique as marker particles for cells. The uptake of the nonfunctionalized polyisoprene nanoparticles in different adherent (HeLa) and also suspension (Jurkat) cell lines is extremely good and fast compared to other polymeric particles, and leads to high-loading efficiencies of the cell (see Figure 8.6). The intracellular polyisoprene particles are localized as single particles in endosomes distributed throughout the entire cytoplasm. When using additionally poly-l-lysine as a transfection agent, the internalization of particles is

8.5.8.4 Uptake Influenced by the Size

In addition to the surface characteristics, particle size plays a crucial role for particle internalization in cells. Rejman et al. evaluated the internalization of PS nanoparticles of various sizes using nonphagocytotic B16 cells [156]. The observed differences of the uptake characteristics were attributed to the occurrence of two uptake mechanisms. For particles of a size of <200 nm a clathrin-mediated mechanism is proposed while a caveolae-mediated mechanism is likely to apply for particles >200 nm. Macropinocytosis—another uptake mechanism—was not excluded in this study. Furthermore Lai et al. have shown that particles of a size of around 20 nm are internalized into cells taking a hitherto unknown clathrin- and caveolae-independent pathway [157].

On the other hand, size may not be the most efficient way of altering the uptake of nanoparticles. In a set of experiments we have combined an emulsion/solvent evaporation method and miniemulsion techniques in formulation of biodegradable nanoparticles using different polymers such as poly(L-lactide) (PLLA), poly(DL-lactide-co-glycolide) (PLGA), and poly (ε-caprolactone) (PCL). Cellular uptake of the obtained particles was investigated in Jurkat and HeLa cells. In the investigated particle size range of 80 to 206 nm, the surfactant on the particles' surface had a greater influence than the particle size (unpublished data). Uptake kinetics reveals that the PLLA and PCL particles are endocytosed much faster than the polystyrene ones. This maybe explained by different uptake mechanism.

8.5.9 Future Directions for Nanoreactors and Mammalian (Stem) Cells

Most nanodevices manufactured today are not performing an active task in the context of mammalian (stem) cells, that is, they are relatively unresponsive to their environment. Once the technology of manufacturing and altering the properties of nanoparticles is fully harnessed they will become more complex and will be adaptive to their environment. We feel that we are just on the edge of combing different technologies so that nanoparticles and nanoreactors can become "smart." One important step is the achievement of nanocapsules as this enables us to have an inside/outside with a shell that can be tailored so that only certain molecules can enter or leave this compartment. This segregation of inside/outside is the most important feature of cells as this enables us to build an ordered entity inside this compartment while the outside is getting less ordered. This would then enable us to include functionalities into these compartments that alter substances entering these nanocapsules, getting metabolized, and then the products could be deposited in an ordered fashion inside the nanocompartment. Furthermore, certain functionalities could only be activated in certain cell types or cell compartments (e.g., depending on pH or other cellular products like hormones, specific metabolic products, or enzymes). Thereby these nanocontainers would be "silent" in nontarget cell types but active in the desired cell type. There they could degrade unwanted cellular products, such as reactive oxygen species that harm cellular functions.

Certainly, these goals will have to be achieved step by step and there will be a long way ahead of us until we will see "nanorobots" that can repair certain cell types in cell cultures or even in vivo.

References

[1] Liew, C.G., et al., "Human embryonic stem cells: possibilities for human cell transplantation." *Ann. Med.*, 2005. 37(7): pp. 521–532.

[2] Cao, F., et al., "Spatial and Temporal Kinetics of Teratoma Formation from Murine Embryonic Stem Cell Transplantation." *Stem Cells Dev.*, 2007.

[3] Shizuru, J.A., R.S. Negrin, and I.L. Weissman, "Hematopoietic stem and progenitor cells: clinical and preclinical regeneration of the hematolymphoid system." *Annu. Rev. Med.*, 2005. 56: pp. 509–538.

[4] Weissman, I.L., D.J. Anderson, and F. Gage, "Stem and progenitor cells: origins, phenotypes, lineage commitments, and transdifferentiations." *Annu. Rev. Cell Dev. Biol.*, 2001. 17: pp. 387–403.

[5] Luzio, J.P., P.R. Pryor, and N.A. Bright, "Lysosomes: fusion and function." *Nat. Rev. Mol. Cell Biol.*, 2007. 8(8): pp. 622–632.

[6] Leibfried, A. and Y. Bellaiche, "Functions of endosomal trafficking in Drosophila epithelial cells." *Curr. Opin. Cell Biol.*, 2007. 19(4): pp. 446–452.

[7] van Dam, E.M. and W. Stoorvogel, "Dynamin-dependent transferrin receptor recycling by endosome-derived clathrin-coated vesicles." *Mol. Biol. Cell*, 2002. 13(1): pp. 169–182.

[8] Miwako, I., T. Schroter, and S.L. Schmid, "Clathrin- and dynamin-dependent coated vesicle formation from isolated plasma membranes." *Traffic*, 2003. 4(6): pp. 376–389.

[9] Pelham, H.R. and J.E. Rothman, "The debate about transport in the Golgi—two sides of the same coin?" *Cell*, 2000. 102(6): pp. 713–719.

[10] Prydz, K. and K.T. Dalen, "Synthesis and sorting of proteoglycans." J. Cell Sci., 2000. 113 Pt 2: pp. 193–205.

[11] Lehmann-Horn, F. and K. Jurkat-Rott, "Voltage-gated ion channels and hereditary disease." *Physiol. Rev.*, 1999. 79(4): pp. 1317–1372.

[44] Le Blanc, K., "Immunomodulatory effects of fetal and adult mesenchymal stem cells." *Cytotherapy*, 2003. 5(6): pp. 485–489.

[45] Ringden, O., et al., "Mesenchymal stem cells for treatment of therapy-resistant graft-versus-host disease." *Transplantation*, 2006. 81(10): pp. 1390–1397.

[46] Lazarus, H.M., et al., "Cotransplantation of HLA-identical sibling culture-expanded mesenchymal stem cells and hematopoietic stem cells in hematologic malignancy patients." *Biol. Blood Marrow Transplant*, 2005. 11(5): pp. 389–398.

[47] Studeny, M., et al., "Mesenchymal stem cells: potential precursors for tumor stroma and targeted-delivery vehicles for anticancer agents." *J. Natl. Cancer Inst.*, 2004. 96(21): pp. 1593–1603.

[48] Schrepfer, S., et al., "Stem cell transplantation: the lung barrier." *Transplant Proc.*, 2007. 39(2): pp. 573–576.

[49] Phinney, D.G., et al., "Murine mesenchymal stem cells transplanted to the central nervous system of neonatal versus adult mice exhibit distinct engraftment kinetics and express receptors that guide neuronal cell migration." *Stem Cells Dev.*, 2006. 15(3): pp. 437–447.

[50] Prockop, D.J. and S.D. Olson, "Clinical trials with adult stem/progenitor cells for tissue repair: let's not overlook some essential precautions." *Blood*, 2007. 109(8): pp. 3147–3151.

[51] Amsalem, Y., et al., "Iron-oxide labeling and outcome of transplanted mesenchymal stem cells in the infarcted myocardium." *Circulation*, 2007. 116(11 Suppl): pp. I38–I45.

[52] Schafer, R., et al., "Transferrin receptor upregulation: in vitro labeling of rat mesenchymal stem cells with superparamagnetic iron oxide." *Radiology*, 2007. 244(2): pp. 514–523.

[53] Braydich-Stolle, L., et al., "In vitro cytotoxicity of nanoparticles in mammalian germline stem cells." *Toxicol. Sci.*, 2005. 88(2): pp. 412–419.

[54] Kostura, L., et al., "Feridex labeling of mesenchymal stem cells inhibits chondrogenesis but not adipogenesis or osteogenesis." *NMR Biomed.*, 2004. 17(7): pp. 513–517.

[55] Arbab, A.S., et al., "Labeling of cells with ferumoxides-protamine sulfate complexes does not inhibit function or differentiation capacity of hematopoietic or mesenchymal stem cells." *NMR Biomed.*, 2005. 18(8): pp. 553–559.

[56] Hsieh, S.C., et al., "The internalized CdSe/ZnS quantum dots impair the chondrogenesis of bone marrow mesenchymal stem cells." *J. Biomed. Mater. Res. B Appl. Biomater.*, 2006. 79(1): pp. 95–101.

[57] Laffon, B., et al., "Individual sensitivity to DNA damage induced by styrene in vitro: influence of cytochrome p450, epoxide hydrolase and glutathione S-transferase genotypes." *Toxicology*, 2003. 186(1-2): pp. 131–141.

[58] Nahar, M., et al., "Functional polymeric nanoparticles: an efficient and promising tool for active delivery of bioactives." *Crit. Rev. Ther. Drug Carrier Syst.*, 2006. 23(4): pp. 259–318.

[59] Weiss, C.K., et al., Cellular uptake behavior of unfunctionalized and functionalized PBCA particles prepared in a miniemulsion. Macromol Biosci, 2007. 7(7): pp. 883–896.

[60] Landfester, K., "Polyreactions in Miniemulsions." *Macromol. Rapid Commun.*, 2001. 22: pp. 896–936.

[61] Landfester, K., "Synthesis of Colloidal Particles in Miniemulsions." *Annual Reviews of Material Research*, 2006. 36: pp. 231–279.

[62] Bender, A.R., et al., "Efficiency of Nanoparticles as a Carrier System for Antiviral Agents in Human Immunodeficiency Virus-Infected Human Monocytes/Macrophages In Vitro." *Antimicrobal Agents and Chemotherapy*, 1996. 40(6): pp. 1467–1471.

[63] Damge, C., et al., "New approach for oral administration of insulin with polyalkylcyanoacrylate nanocapsules as drug carrier." *Diabetes*, 1988. 37(2): pp. 246–251.

[64] Jiao, Y., et al., "In Vitro and In Vivo Evaluation of Oral Heparin-Loaded Polymeric Nanoparticles in Rabbits." *Circulation*, 2002. 105(2): pp. 230–235.

[65] Vauthier, C., et al., "Poly(alkylcyanoacrylates) as biodegradable materials for biomedical applications." *Advanced Drug Delivery Reviews*, 2003. 55: pp. 519–548.

[66] Kreuter, J., "Nanoparticulate systems for brain delivery of drugs." *Advanced Drug Delivery Reviews*, 2001. 47: pp. 65–81.

[67] Gref, R., et al., "Biodegradable Long-Circulating Polymeric Nanospheres." *Science*, 1994. 263: pp. 1600–1603.

[68] Gurny, R., et al., "Development of biodegradable and injectable latices for controlled release of potent drugs." *Drug Development and Industrial Pharmacy*, 1981. 7(1): pp. 1–25.

[69] Vonarbourg, A., et al., "Parameters influencing the stealthiness of colloidal drug delivery systems." *Biomaterials*, 2006. 27: pp. 4356–4373.

[70] Torchilin, V.P. and V.S. Trubetskoy, "Which polymers can make nanoparticulate drug carriers long-circulating?" *Advanced Drug Delivery Reviews*, 1995. 16: pp. 141–155.

[71] Schauer, R., "Sialic acids and their role as biological masks." *Trends in Biochemical Sciences*, 1985. 10(9): pp. 357–360.

[72] Passirani, C., et al., "Long-Circulating Nanoparticles Bearing Heparin or Dextran Covalently Bound to Poly(Methyl Methacrylate)." *Pharmaceutical Research*, 1998. 15(7): pp. 1046–1050.

[102] Hunter, A.C., "Molecular hurdles in polyfectin design and mechanistic background to polycation induced cytotoxicity." *Adv. Drug Deliv. Rev.*, 2006. 58(14): pp. 1523–1531.

[103] Symonds, P., et al., "Low and high molecular weight poly(L-lysine)s/poly(L-lysine)-DNA complexes initiate mitochondrial-mediated apoptosis differently." *FEBS Lett.*, 2005. 579(27): pp. 6191–6198.

[104] Zimmet, J. and J. Hare, "Emerging role for bone marrow derived mesenchymal stem cells in myocardial regenerative therapy." *Basic Res. Cardiol.*, 2005. 100(6): pp. 471–481.

[105] Kemp, K., J. Hows, and C. Donaldson, "Bone marrow-derived mesenchymal stem cells." *Leuk. Lymphoma*, 2005. 46(11): pp. 1531–1544.

[106] Noel, D., F. Djouad, and C. Jorgense, "Regenerative medicine through mesenchymal stem cells for bone and cartilage repair." *Curr. Opin. Investig. Drugs*, 2002. 3(7): pp. 1000–1004.

[107] Holzapfel, V., et al., Preparation of Fluorescent Carboxyl and Amino Functionalized Polystyrene Particles by Minie-mulsion Polymerization as Markers for Cells." *J. Phys.: Condens. Matter*, 2006. 18: pp. S2581–S2594.

[108] Aime, S., L. Frullano, and S. Geninatti Crich, "Compartmentalization of a gadolinium complex in the apoferritin cavity: a route to obtain high relaxivity contrast agents for magnetic resonance imaging." *Angew Chem. Int. Ed. Engl.*, 2002. 41(6): pp. 1017–1019.

[109] Morawski, A.M., et al., "Targeted nanoparticles for quantitative imaging of sparse molecular epitopes with MRI." *Magn. Reson. Med.*, 2004. 51(3): pp. 480–486.

[110] Reynolds, C.H., et al., *Gadolinium-Loaded Nanoparticles: New Contrast Agents for Magnetic Resonance Imaging.* 2000. pp. 8940–8945.

[111] Toth, E.E., et al., "Water Exchange and Rotational Dynamics of the Dimeric Gadolinium(III) Complex [BO{Gd(DO3A)(H(2)O)}(2)]: A Variable-Temperature and -Pressure (17)O NMR Study(1)." *Inorg. Chem.*, 1996. 35(11): pp. 3375–3379.

[112] Jagielski, N., et al., "Nanocapsules synthesized by Miniemulsion Technique for Application as New Contrast Agent Materials." *Macromol. Chem. Phys.*, 2007. 208(19-20): pp. 2229–2241.

[113] Hattori, Y. and Y. Maitani, "Enhanced in vitro DNA transfection efficiency by novel folate-linked nanoparticles in human prostate cancer and oral cancer." *J. Control Release*, 2004. 97(1): pp. 173–183.

[114] Sahoo, S.K. and V. Labhasetwar, Enhanced antiproliferative activity of transferrin-conjugated paclitaxel-loaded nanoparticles is mediated via sustained intracellular drug retention." *Mol. Pharm.*, 2005. 2(5): pp. 373–383.

[115] Mullis, K., et al., "Specific enzymatic amplification of DNA in vitro: the polymerase chain reaction." *Cold Spring Harb. Symp. Quant. Biol.*, 1986. 51 Pt 1: pp. 263–273.

[116] Falsey, A.R., M.A. Formica, and E.E. Walsh, "Diagnosis of respiratory syncytial virus infection: comparison of reverse transcription-PCR to viral culture and serology in adults with respiratory illness." *J. Clin. Microbiol.*, 2002. 40(3): pp. 817–820.

[117] Ghadessy, F.J., J.L. Ong, and P. Holliger, "Directed evolution of polymerase function by compartmentalized self-replication." *Proc. Natl. Acad. Sci. USA*, 2001. 98(8): pp. 4552–4557.

[118] Nakano, M., et al., "Single-molecule PCR using water-in-oil emulsion." *J. Biotechnol.*, 2003. 102(2): pp. 117–124.

[119] Dressman, D., et al., "Transforming single DNA molecules into fluorescent magnetic particles for detection and enumeration of genetic variations." *Proc. Natl. Acad. Sci. USA*, 2003. 100(15): pp. 8817–8822.

[120] Tawfik, D.S. and A.D. Griffiths, "Man-made cell-like compartments for molecular evolution." *Nat. Biotechnol.*, 1998. 16(7): pp. 652–656.

[121] Musyanovych, A., V. Mailander, and K. Landfester, "Miniemulsion droplets as single molecule nanoreactors for polymerase chain reaction." *Biomacromolecules*, 2005. 6(4): pp. 1824–1828.

[122] Piatek, A.S., et al., "Molecular beacon sequence analysis for detecting drug resistance in Mycobacterium tuberculosis." *Nat. Biotechnol.*, 1998. 16(4): pp. 359–363.

[123] Pierce, G.F., et al., "Gene therapy, bioengineered clotting factors and novel technologies for hemophilia treatment." *J. Thromb. Haemost.*, 2007. 5(5): pp. 901–906.

[124] Hacein-Bey-Abina, S., et al., "LMO2-associated clonal T cell proliferation in two patients after gene therapy for SCID-X1." *Science*, 2003. 302(5644): pp. 415–419.

[125] Sokolova, V.V., et al., "Effective transfection of cells with multi-shell calcium phosphate-DNA nanoparticles." *Biomaterials*, 2006. 27(16): pp. 3147–3153.

[126] Guo, C. and R.A. Gemeinhart, "Assessment of a modular transfection system based upon cellular localization of DNA." *Mol. Pharm.*, 2004. 1(4): pp. 309–316.

[127] Luo, D. and W.M. Saltzman, "Enhancement of transfection by physical concentration of DNA at the cell surface." *Nat. Biotechnol.*, 2000. 18(8): pp. 893–895.

[128] Arnida, et al., "PEGylated gene nanocarriers based on block catiomers bearing ethylenediamine repeating units directed to remarkable enhancement of photochemical transfection." *J. Control Release*, 2006. 115(2): pp. 208–215.

[129] Berger, C., et al., "Feasibility and limits of magnetically labeling primary cultured rat T cells with ferumoxides coupled with commonly used transfection agents." *Mol. Imaging*, 2006. 5(2): pp. 93–104.

[130] Wiehe, J.M., et al., "mRNA-Mediated Gene Delivery into Human Progenitor Cells Promotes Highly Efficient Protein Expression." *J. Cell Mol. Med.*, 2007. 11(3): pp. 521–530.

About the Editors

Agnes Ostafin joined the faculty of Materials Science & Engineering at the University of Utah in 2006 as an associate professor, and holds an adjunct appoint in the Department of Bioengineering. After receiving bachelor degrees in biological sciences and chemistry from Wayne State University, Detroit, Michigan, she received her doctoral degree in physical chemistry from the University of Minnesota, Twin Cities, after working with Professor S. Lipsky on photophysical processes in hydrocarbon materials. She completed two years of postdoctoral training in the Chemistry Division at Argonne National Laboratory working on electron transfer processes in natural photosynthetic materials, and another two years at the University of Chicago studying genetic engineering of photosynthetic prokaryotes, and structure-function relationships in photosynthetic proteins. Her research program involves the development of inorganic nanoreactors for biosensing and optical applications, along with studies of the properties of biomembranes and their interactions with nanomaterials. She has several patents/patent applications in the area of nanoparticle/nanoreactor design and utility, and has served on numerous review panels and professional committees. Most recently she served as treasurer of the Institute of Biological Engineering.

Katharina Landfester joined the Max Planck Society in 2008 as one of the directors of the Max Planck Institute for Polymer Research. She studied chemistry at the Technical University of Darmstadt. For her diploma thesis, she was at the Ecole d'Application des Hautes PolymPres in Strasbourg (Professo M. Lambla). In 1995, she received her doctoral degree in physical

I.M. Okhapkin
Nesmeyanov Institute of Organoelement Compounds
Vavilova srt. 28
Moscow, 119991, Russia

Agnes Ostafin
University of Utah
Material Science and Engineering
122 S. Central Campus Drive
Salt Lake City, UT 84112

Owe Orwar
Department of Chemical and Biological Engineering
Microtechnology Center
Chalmers University of Technology
Kemiv en 10SE-41296 Geborg, Sweden

Hartley Schmidt
Intel
Albequerque, NM 87144

Hubert Schrenzenmeier
University of Ulm
Institute for Clinical Transfusion Medicine and Immunogenetics
Institute for Transfusion Medicine
University of Ulm
Helmholtzstr. 10
89081 Ulm

V.V. Vasilevskaya
Nesmeyanov Institute of Organoelement Compounds
Vavilova srt. 28
Moscow, 119991, Russia

Qiang Wang
Haliburton
Duncan, OK 73533

Philip Wingert
Rice University
Department of Bioengineering
Houston, TX 77251

Index